THE NUMBER-SYSTEM OF ALGEBRA

TREATED THEORETICALLY AND HISTORICALLY

HENRY B. FINE, PH. D.
PROFESSOR OF MATHEMATICS IN PRINCETON UNIVERSITY

HAWK PRESS

Published by

Hawk Press
4836/24, Ansari Road, Daryaganj
New Delhi - 110 002
Phones : 9643330713, + 91-11-23278618, + 91-11-35676207
E-mail : thehawkpress@gmail.com
www.thehawkpress.com

ISBN : 978-93-93971-66-1

THE NUMBER-SYSTEM OF ALGEBRA

PREFACE.

The theoretical part of this little book is an elementary exposition of the nature of the number concept, of the positive integer, and of the four artificial forms of number which, with the positive integer, constitute the "number-system" of algebra, viz. the negative, the fraction, the irrational, and the imaginary. The discussion of the artificial numbers follows, in general, the same lines as my pamphlet: *On the Forms of Number arising in Common Algebra*, but it is much more exhaustive and thorough-going. The point of view is the one first suggested by Peacock and Gregory, and accepted by mathematicians generally since the discovery of quaternions and the Ausdehnungslehre of Grassmann, that algebra is completely defined formally by the laws of combination to which its fundamental operations are subject; that, speaking generally, these laws alone define the operations, and the operations the various artificial numbers, as their formal or symbolic results. This doctrine was fully developed for the negative, the fraction, and the imaginary by Hankel, in his *Complexe Zahlensystemen*, in 1867, and made complete by Cantor's beautiful theory of the irrational in 1871, but it has not as yet received adequate treatment in English.

Any large degree of originality in work of this kind is naturally out of the question. I have borrowed from a great many sources, especially from Peacock, Grassmann, Hankel, Weierstrass, Cantor, and Thomae (*Theorie der analytischen Functionen einer complexen Veränderlichen*). I may mention, however, as more or less distinctive features of my discussion, the treatment of number, counting (§§ 1–5), and the equation (§§ 4, 12), and the prominence given the laws of the determinateness of subtraction and division.

Much care and labor have been expended on the historical chapters of the book. These were meant at the outset to contain only a brief account of the origin and history of the artificial numbers. But I could not bring myself to ignore primitive counting and the development of numeral notation, and I soon found that a clear and connected account of the origin of the negative and imaginary is possible only when embodied in a sketch of the early history of the equation. I have thus been led to write a *résumé* of the history of the most important parts of elementary arithmetic and algebra.

Moritz Cantor's *Vorlesungen über die Geschichte der Mathematik*, Vol. I, has been my principal authority for the entire period which it covers, *i. e.* to 1200 A. D. For the little I have to say on the period 1200 to 1600, I have depended chiefly, though by no means absolutely, on Hankel: *Zur Geschichte der Mathematik in Altertum und Mittelalter.* The remainder of my sketch is for the most part based on the original sources.

HENRY B. FINE.

PRINCETON, April, 1891.

In this second edition a number of important corrections have been made. But there has been no attempt at a complete revision of the book.

HENRY B. FINE.

PRINCETON, September, 1902.

Contents

PRINCIPAL FOOTNOTES

Part I

THEORETICAL

1. THE POSITIVE INTEGER, AND THE LAWS WHICH REGULATE THE ADDITION AND MULTIPLICATION OF POSITIVE INTEGERS.

1. Number. We say of certain distinct things that they form a group[1] when we make them collectively a single object of our attention.

The *number of things* in a group is that property of the group which remains unchanged during every change in the group which does not destroy the separateness of the things from one another or their common separateness from all other things.

Such changes may be changes in the characteristics of the things or in their arrangement within the group. Again, changes of arrangement may be changes either in the order of the things or in the manner in which they are associated with one another in smaller groups.

We may therefore say:

The number of things in any group of distinct things is independent of the characters of these things, of the order in which they may be arranged in the group, and of the manner in which they may be associated with one another in smaller groups.

2. Numerical Equality. The number of things in any two groups of distinct things is the same, when for each thing in the first group there is one in the second, and reciprocally, for each thing in the second group, one in the first.

Thus, the number of letters in the two groups, A, B, C; D, E, F, is the same. In the second group there is a letter which may be assigned to each of the letters in the first: as D to A, E to B, F to C; and reciprocally, a letter in the first which may be assigned to each in the second: as A to D, B to E, C to F.

Two groups thus related are said to be in *one-to-one* (1–1) *correspondence.*

Underlying the statement just made is the assumption that if the two groups correspond in the manner described for one order of the things in each, they will correspond if the things be taken in any other order also; thus, in the example given, that if E instead of D be assigned to A, there will again be a letter in the group D, E, F, viz. D or F, for each of the remaining letters B and C, and reciprocally. This is an immediate consequence of § 1, foot-note.

The number of things in the first group is *greater than* that in the second, or the number of things in the second *less than* that in the first, when there is one thing in the first group for each thing in the second, but *not* reciprocally one in the second for each in the first.

3. Numeral Symbols. As regards the number of things which it contains, therefore, a group may be represented by any other group, *e. g.* of the fingers or of simple marks, |'s, which stands to it in the relation of correspondence described in § 2. This is the primitive method of representing the number of things in a group and, like the modern method, makes it possible to compare numerically groups which are separated in time or space.

The modern method of representing the number of things in a group differs from the primitive only in the substitution of symbols, as 1, 2, 3, etc., or numeral words, as *one, two, three*, etc., for the various groups of marks |, ||, |||, etc. These symbols are the positive integers of arithmetic.

[1] By group we mean *finite* group, that is, one which cannot be brought into one-to-one correspondence (§ 2) with any part of itself.

A positive integer is a symbol for the number of things in a group of distinct things.

For convenience we shall call the positive integer which represents the number of things in any group its numeral symbol, or when not likely to cause confusion, its number simply,—this being, in fact, the primary use of the word "number" in arithmetic.

In the following discussion, for the sake of giving our statements a general form, we shall represent these numeral symbols by letters, a, b, c, etc.

4. The Equation. The numeral symbols of two groups being a and b; when the number of things in the groups is the same, this relation is expressed by the *equation*

$$a = b;$$

when the first group is greater than the second, by the *inequality*

$$a > b;$$

when the first group is less than the second, by the *inequality*

$$a < b.$$

A numerical equation is thus a declaration in terms of the numeral symbols of two groups and the symbol = *that these groups are in one-to-one correspondence* (§2).

5. Counting. The fundamental operation of arithmetic is counting.

To count a group is to set up a one-to-one correspondence between the individuals of this group and the individuals of some representative group.

Counting leads to an expression for the number of things in any group in terms of the representative group: if the representative group be the fingers, to a group of fingers; if marks, to a group of marks; if the numeral words or symbols in common use, to one of these words or symbols.

There is a difference between counting with numeral words and the earlier methods of counting, due to the fact that the numeral words have a certain recognized order. As in finger-counting one finger is attached to each thing counted, so here one word; but that word represents numerically not the thing to which it is attached, but the entire group of which this is the last. The same sort of counting may be done on the fingers when there is an agreement as to the order in which the fingers are to be used; thus if it were understood that the fingers were always to be taken in normal order from thumb to little finger, the little finger would be as good a symbol for 5 as the entire hand.

6. Addition. If two or more groups of things be brought together so as to form a single group, the numeral symbol of this group is called the *sum* of the numbers of the separate groups.

If the sum be s, and the numbers of the separate groups a, b, c, etc., respectively, the relation between them is symbolically expressed by the equation

$$s = a + b + c + \text{etc.},$$

where the sum-group is supposed to be formed by joining the second group—to which b belongs—to the first, the third group—to which c belongs—to the resulting group, and so on.

The operation of finding s when a, b, c, etc., are known, is *addition*.

Addition is abbreviated counting.

Addition is subject to the two following laws, called the *commutative* and *associative* laws respectively, viz.:

$$\text{I.}\quad a + b = b + a.$$
$$\text{II.}\quad a + (b + c) = a + b + c.$$

Or,

I. To add b to a is the same as to add a to b.

II. To add the sum of b and c to a is the same as to add c to the sum of a and b.

Both these laws are immediate consequences of the fact that the sum-group will consist of the same individual things, and the number of things in it therefore be the same, whatever the order or the combinations in which the separate groups are brought together (§1).

7. Multiplication. The sum of b numbers each of which is a is called the *product* of a by b, and is written $a \times b$, or $a \cdot b$, or simply ab.

The operation by which the product of a by b is found, when a and b are known, is called *multiplication*.

Multiplication is an abbreviated addition.

Multiplication is subject to the three following laws, called respectively the *commutative, associative*, and *distributive* laws for multiplication, viz.:

III. $ab = ba$.

IV. $a(bc) = abc$.

V. $a(b + c) = ab + ac$.

Or,

III. The product of a by b is the same as the product of b by a.

IV. The product of a by bc is the same as the product of ab by c.

V. The product of a by the sum of b and c is the same as the sum of the product of a by b and o

These laws are consequences of the commutative and associative laws for addition.

Thus,

III. *The Commutative Law.* The units of the group which corresponds to the sum of b numbers each equal to a may be arranged in b rows containing a units each. But in such an arrangement there are a columns containing b units each; so that if this same set of units be grouped by columns instead of rows, the sum becomes that of a numbers each equal to b, or ba. Therefore $ab = ba$, by the commutative and associative laws for addition.

IV. *The Associative Law.*

$$\begin{aligned} abc &= c \text{ sums such as } (a + a + \cdots \text{ to } b \text{ terms}) \\ &= a + a + a + \cdots \text{ to } bc \text{ terms (by the associative law for addition)} \\ &= a(bc). \end{aligned}$$

V. *The Distributive Law.*

$$\begin{aligned} a(b + c) &= a + a + a + \cdots \text{to } (b + c) \text{ terms} \\ &= a + a + \cdots \text{ to } b \text{ terms}) + (a + a + \cdots \text{ to } c \text{ terms}) \\ &\qquad \text{(by the associative law for addition),} \\ &= ab + ac. \end{aligned}$$

The commutative, associative, and distributive laws for sums of any number of terms and products of any number of factors follow immediately from I–V. Thus the product of the factors a, b, c, d, taken in any two orders, is the same, since the one order can be transformed into the other by successive interchanges of consecutive letters.

2. SUBTRACTION AND THE NEGATIVE INTEGER.

8. Numerical Subtraction. Corresponding to every mathematical operation there is another, commonly called its *inverse*, which exactly undoes what the operation itself does. Subtraction stands in this relation to addition, and division to multiplication.

To *subtract* b from a is to find a number to which if b be added, the sum will be a. The result is written $a - b$; by definition, it identically satisfies the equation

VI. $(a - b) + b = a$;

that is to say, $a - b$ is the number belonging to the group which with the b-group makes up the a-group.

Obviously subtraction is always possible when b is less than a, but then only. Unlike addition, in each application of this operation regard must be had to the relative size of the two numbers concerned.

9. Determinateness of Numerical Subtraction. Subtraction, when possible, is a *determinate* operation. There is but *one* number which will satisfy the equation $x + b = a$, but one number the sum of which and b is a. In other words, $a - b$ is one-valued.

For if c and d both satisfy the equation $x + b = a$, since then $c + b = a$ and $d + b = a$, $c + b = d + b$; that is, a one-to-one correspondence may be set up between the individuals of the $(c+b)$ and $(d+b)$ groups (§4). The same sort of correspondence, however, exists between any b individuals of the first group and any b individuals of the second; it must, therefore, exist between the remaining c of the first and the remaining d of the second, or $c = d$.

This characteristic of subtraction is of the same order of importance as the commutative and associative laws, and we shall add to the group of laws I–V and definition VI—as being, like them, a fundamental principle in the following discussion—the theorem

VII. $\begin{cases} \text{If } a + c &= b + c \\ \quad\quad\; a &= b, \end{cases}$

which may also be stated in the form: If one term of a sum changes while the other remains constant, the sum changes. The same reasoning proves, also, that

VIII. $\begin{cases} \text{As } a + c > & \text{or} \quad < b + c \\ \quad\quad\quad\; a & \text{or} \quad b, \end{cases}$

10. Formal Rules of Subtraction. All the rules of subtraction are purely *formal* consequences of the fundamental laws I–V, VII, and definition VI. They must follow, whatever the meaning of the symbols a, b, c, $+$, $-$, $=$; a fact which has an important bearing on the following discussion.

It will be sufficient to consider the equations which follow. For, properly combined, they determine the result of any series of subtractions or of any complex operation made up of additions, subtractions, and multiplications.

1. $a - (b + c) = a - b - c = a - c - b.$
2. $a - (b - c) = a - b + c.$
3. $a + b - b = a.$
4. $a + (b - c) = a + b - c = a - c + b.$
5. $a(b - c) = ab - ac.$

For

1. $a - b - c$ is the form to which if first c and then b be added; or, what is the same thing (by I), first b and then c; or, what is again the same thing (by II), $b+c$ at once,—the sum produced is a (by VI). $a-b-c$ is therefore the same as $a-c-b$, which is as it stands the form to which if b, then c, be added the sum is a; also the same as $a-(b+c)$, which is the form to which if $b+c$ be added the sum is a.

2.
$$\begin{aligned} a-(b-c) &= a-(b-c)-c+c, && \text{Def. VI.}\\ &= a-(b-c+c)+c, && \text{Eq. 1.}\\ &= a-b-c. && \text{Def. VI.}\end{aligned}$$

3.
$$\begin{aligned} a+b-b+b &= a+b. && \text{Def. VI.}\\ \text{But}\quad a+b &= a+b.\\ \therefore a+b-b &= a. && \text{Law VII.}\end{aligned}$$

4.
$$\begin{aligned} a+b-c &= a+(b-c+c)-c, && \text{Def. VI.}\\ &= a+(b-c). && \text{Law II, Eq. 3.}\end{aligned}$$

5.
$$\begin{aligned} ab-ac &= a(b-c+c)-ac, && \text{Def. VI.}\\ &= a(b-c)+ac-ac, && \text{Law V.}\\ &= a(b-c). && \text{Eq. 3.}\end{aligned}$$

Equation 3 is particularly interesting in that it defines addition as the inverse of subtraction. Equation 1 declares that two consecutive subtractions may change places, are commutative. Equations 1, 2, 4 together supplement law II, constituting with it a complete associative law of addition and subtraction; and equation 5 in like manner supplements law V.

11. Limitations of Numerical Subtraction. Judged by the equations 1–5, subtraction is the exact counterpart of addition. It conforms to the same general laws as that operation, and the two could with fairness be made to interchange their rôles of direct and inverse operation.

But this equality proves to be only apparent when we attempt to interpret these equations. The requirement that subtrahend be less than minuend then becomes a serious restriction. It makes the range of subtraction much narrower than that of addition. It renders the equations 1–5 available for special classes of values of a, b, c only. If it must be insisted on, even so simple an inference as that $a-(a+b)+2b$ is equal to b cannot be drawn, and the use of subtraction in any reckoning with symbols whose relative values are not at all times known must be pronounced unwarranted.

One is thus naturally led to ask whether to be valid an algebraic reckoning must be interpretable numerically and, if not, to seek to free subtraction and the rules of reckoning with the results of subtraction from a restriction which we have found to be so serious.

12. Symbolic Equations. Principle of Permanence. Symbolic Subtraction. In pursuance of this inquiry one turns first to the equation $(a-b)+b=a$, which serves as a definition of subtraction when b is less than a.

This is an equation in the primary sense (§ 4) only when $a-b$ is a number. But in the broader sense, that

An equation is any declaration of the equivalence of definite combinations of symbols—equivalence in the sense that one may be substituted for the other,— $(a - b) + b = a$ may be an equation, whatever the values of a and b.

And if no different meaning has been attached to $a-b$, and it is declared that $a-b$ is the symbol which associated with b in the combination $(a - b) + b$ is equivalent to a, this declaration, or the *equation*

$$(a - b) + b = a,$$

is a *definition*[1] of this symbol.

By the assumption of the *permanence of form* of the numerical equation in which the definition of subtraction resulted, one is thus put immediately in possession of a *symbolic* definition of subtraction which is general.

The numerical definition is subordinate to the symbolic definition, being the interpretation of which it admits when b is less than a.

But from the standpoint of the symbolic definition, interpretability—the question whether $a - b$ is a number or not—is irrelevant; only such properties may be attached to $a - b$, by itself considered, as flow immediately from the generalized equation

$$(a - b) + b = a.$$

In like manner each of the fundamental laws I–V, VII, on the assumption of the *permanence of its form* after it has ceased to be interpretable numerically, becomes a declaration of the equivalence of certain definite combinations of symbols, and the formal consequences of these laws—the equations 1–5 of § 10—become definitions of addition, subtraction, multiplication, and their mutual relations—definitions which are purely symbolic, it may be, but which are unrestricted in their application.

These definitions are legitimate from a logical point of view. For they are merely the laws I–VII, and we may assume that these laws are *mutually consistent* since we have proved that they hold good for positive integers. Hence, if *used correctly*, there is no more possibility of their leading to false results than there is of the more tangible numerical definitions leading to false results. The laws of correct thinking are as applicable to mere symbols as to numbers.

What the value of these symbolic definitions is, to what extent they add to the power to draw inferences concerning numbers, the elementary algebra abundantly illustrates.

One of their immediate consequences is the introduction into algebra of two new symbols, *zero* and the *negative*, which contribute greatly to increase the simplicity, comprehensiveness, and power of its operations.

13. Zero. When b is set equal to a in the general equation

$$(a - b) + b = a,$$

it takes one of the forms

$$(a - a) + a = a,$$

$$(b - b) + b = b.$$

It may be proved that

[1] A definition in terms of symbolic, not numerical addition. The sign + can, of course, indicate numerical addition only when both the symbols which it connects are numbers.

$$
\begin{aligned}
& & a - a &= b - b. & \\
&\text{For} & (a - a) + (a + b) &= (a - a) + a + b, & \text{Law II.} \\
& & &= a + b, & \\
&\text{since} & (a - a) + a &= a. & \\
&\text{And} & (b - b) + (a + b) &= (b - b) + b + a, & \text{Laws I, II.} \\
& & &= b + a, & \\
&\text{since} & (b - b) + b &= b. & \\
&\text{Therefore} & a - a &= b - b. & \text{Law VII.}
\end{aligned}
$$

$a - a$ is therefore altogether independent of a and may properly be represented by a symbol unrelated to a. The symbol which has been chosen for it is 0, called *zero*.

Addition is defined for this symbol by the equations

1

$$
\begin{aligned}
0 + a &= a, & \text{definition of 0.} \\
a + 0 &= a. & \text{Law I.}
\end{aligned}
$$

Subtraction (partially), by the equation

2.

$$
\begin{aligned}
& & a - 0 &= a. & \\
&\text{For} & (a - 0) + 0 &= a. & \text{Def. VI.}
\end{aligned}
$$

Multiplication (partially), by the equations

3.

$$
\begin{aligned}
& & a \times 0 &= 0 \times a = 0. & \\
&\text{For} & a \times 0 &= a(b - b), & \text{definition of 0.} \\
& & &= ab - ab, & \text{§ 10, 5.} \\
& & &= 0. & \text{definition of 0.}
\end{aligned}
$$

14. The Negative. When b is greater than a, equal say to $a+d$, so that $b-a = d$, then

$$
\begin{aligned}
a - b &= a - (a + d), & \\
&= a - a - d, & \text{§ 10, 1.} \\
&= 0 - d. & \text{definition of 0.}
\end{aligned}
$$

For $0 - d$ the briefer symbol $-d$ has been substituted; with propriety, certainly, in view of the lack of significance of 0 in relation to addition and subtraction. The equation $0-d = -d$, moreover, supplies the missing rule of subtraction for 0. (Compare § 13, 2.)

The symbol $-d$ is called the *negative*, and in opposition to it, the number d is called *positive*.

Though in its origin a sign of operation (subtraction from 0), the sign $-$ is here to be regarded merely as part of the symbol $-d$.

$-d$ is as serviceable a substitute for $a-b$ when $a < b$, as is a single numeral symbol when $a > b$.

The rules for reckoning with the new symbol—definitions of its addition, subtraction, multiplication—are readily deduced from the laws I–V, VII, definition VI, and the equations 1–5 of § 10, as follows:

1.

$$
\begin{aligned}
& & b + (-b) &= -b + b = 0. & \\
&\text{For} & -b + b &= (0 - b) + b, & \text{definition of negative.} \\
& & &= 0. & \text{Def. VI.}
\end{aligned}
$$

$-b$ may therefore be defined as the symbol the sum of which and b is 0.

2.

$$\begin{aligned} a+(-b) &= -b+a = a-b. \\ \text{For}\quad a+(-b) &= a+(0-b), && \text{definition of negative.} \\ &= a+0-b, && \S\ 10,\ 4. \\ &= a-b. && \S\ 13,\ 1. \end{aligned}$$

3.

$$\begin{aligned} -a+(-b) &= -(a+b). \\ \text{For}\quad -a+(-b) &= 0-a-b, && \text{by the reasoning in } \S\ 14,\ 2. \\ &= 0-(a+b), && \S 10,1. \\ &= -(a+b). && \text{definition of negative.} \end{aligned}$$

4.

$$\begin{aligned} a-(-b) &= a+b. \\ \text{For}\quad a-(-b) &= a-(0-b), && \text{definition of negative.} \\ &= a-0+b, && \S\ 10,\ 2. \\ &= a+b. && \S 13,\ 2. \end{aligned}$$

5.

$$\begin{aligned} (-a)-(-b) &= b-a. \\ \text{For}\quad -a-(-b) &= -a+b, && \text{by the reasoning in } \S\ 14,\ 4. \\ &= b-a. && \S 14,\ 2. \\ \text{COR.}\quad (-a)-(-a) &= 0. \end{aligned}$$

6.

$$\begin{aligned} a(-b) &= (-b)a = -ab. \\ \text{For}\quad 0 &= a(b-b), && \S 13,\ 3. \\ &= ab+a(-b). && \text{Law V.} \\ \therefore a(-b) &= -ab. && \S\ 14,\ 1;\ \text{Law VII.} \end{aligned}$$

7.

$$\begin{aligned} (-a)\times 0 &= 0\times(-a) = 0. \\ \text{For}\quad (-a)\times 0 &= (-a)(b-b), && \text{definition of 0.} \\ &= (-a)b-(-a)b, && \S\ 10,5. \\ &= 0. && \S\ 14,6,\text{ and } 5,\text{ Cor.} \end{aligned}$$

8.

$$\begin{aligned} (-a)(-b) &= ab. \\ \text{For}\quad 0 &= (-a)(b-b), && \S\ 14,7. \\ &= (-a)b+(-a)(-b), && \text{Law V.} \\ &= -ab+(-a)(-b). && \S\ 14,6. \\ \therefore (-a)(-b) &= ab. && \S\ 14,1;\text{Law VII.} \end{aligned}$$

By this method one is led, also, to definitions of *equality* and greater or lesser *inequality* of negatives. Thus

9.

$$\begin{aligned} & -a >, && = \text{ or } < -b, \\ \text{according as}\quad & b >, && = \text{ or } < a.^{1} \\ \text{For as}\quad & b >, && =, < a, \\ & -a+a+b >, && =, < -b+b+a, && \S\ 14,1;\S\ 13,1. \\ \text{or}\quad & -a >, && =, < -b, && \text{Law VII or VII}'. \\ \text{In like manner}\quad & -a && < 0 < b. \end{aligned}$$

15. Recapitulation. The nature of the argument which has been developed in the present chapter should be carefully observed.

From the definitions of the positive integer, addition, and subtraction, the associative and commutative laws and the determinateness of subtraction followed. The assumption of the permanence of the result $a - b$, as defined by $(a - b) + b = a$, for all values of a and b, led to definitions of the two symbols 0, $-d$, zero and the negative; and from the assumption of the permanence of the laws I–V, VII were derived definitions of the addition, subtraction, and multiplication of these symbols,—the assumptions being just sufficient to determine the meanings of these operations unambiguously.

In the case of numbers, the laws I–V, VII, and definition VI were deduced from the characteristics of numbers and the definitions of their operations; in the case of the symbols 0, $-d$, on the other hand, the characteristics of these symbols and the definitions of their operations were deduced from the laws.

With the acceptance of the negative the character of arithmetic undergoes a radical change.[2] It was already in a sense symbolic, expressed itself in equations and inequalities, and investigated the results of certain operations. But its symbols, equations, and operations were all interpretable in terms of the reality which gave rise to it, the number of things in actually existing groups of things. Its connection with this reality was as immediate as that of the elementary geometry with actually existing space relations.

But the negative severs this connection. The negative is a symbol for the result of an operation which cannot be effected with actually existing groups of things, which is, therefore, purely symbolic. And not only do the fundamental operations and the symbols on which they are performed lose reality; the equation, the fundamental judgment in all mathematical reasoning, suffers the same loss. From being a declaration that two groups of things are in one-to-one correspondence, it becomes a mere declaration regarding two combinations of symbols, that in any reckoning one may be substituted for the other.

[1] On the other hand, $-a$ is said to be *numerically* greater than, equal to, or less than $-b$, according as a is itself greater than, equal to, or less than b.

[2] In this connection see § 25.

3. DIVISION AND THE FRACTION.

16. Numerical Division. The inverse operation to multiplication is division.

To divide a by b is to find a number which multiplied by b produces a. The result is called the quotient of a by b, and is written $\frac{a}{b}$. By definition

$$\left(\frac{a}{b}\right) b = a$$

Like subtraction, division cannot be always effected. Only in exceptional cases can the a-group be subdivided into b equal groups.

17. Determinateness of Numerical Division. When division can be effected at all, it can lead to but a single result; it is *determinate.*

For there can be but one number the product of which by b is a; in other words,

$$\left\{\begin{array}{ll} \text{If} \quad cb &= db, \\ \qquad c &= d.^1 \end{array}\right.$$

For b groups each containing c individuals cannot be equal to b groups each containing d individuals unless $c = d$ (§4).

This is a theorem of fundamental importance. It may be called the law of determinateness of division. It declares that if a product and one of its factors be determined, the remaining factor is definitely determined also; or that if one of the factors of a product changes while the other remains unchanged, the product changes. It alone makes division in the arithmetical sense possible. The fact that it does not hold for the symbol 0, but that rather a product remains unchanged (being always 0) when one of its factors is 0, however the other factor be changed, makes division by 0 impossible, rendering unjustifiable the conclusions which can be drawn in the case of other divisors.

The reasoning which proved law IX proves also that

$$\text{IX'}. \qquad \left\{\begin{array}{lll} \text{As} \quad cb > & \text{or} & < db, \\ \qquad c > & \text{or} & < d. \end{array}\right.$$

18. Formal Rules of Division. The fundamental laws of the multiplication of numbers are

$$\begin{array}{lrl} \text{III.} & ab & = ba, \\ \text{IV.} & a(bc) & = abc, \\ \text{V.} & a(b+c) & = ab + ac. \end{array}$$

Of these, the definition

$$\text{VIII.} \qquad \left(\frac{a}{b}\right) b = a,$$

the theorem

$$\text{IX.} \qquad \left\{\begin{array}{lll} \text{If } ac & = bc, & \\ \qquad a & = b, & \text{unless } c = 0, \end{array}\right.$$

and the corresponding laws of addition and subtraction, the rules of division are purely *formal* consequences, deducible precisely as the rules of subtraction 1–5 of §10

[1] The case $b = 0$ is excluded, 0 not being a number in the sense in which that word is here used.

in the preceding chapter. They follow without regard to the meaning of the symbols a, b, c, $=$, $+$, $-$, ab, $\frac{a}{b}$. Thus:

1.
$$\frac{a}{b} \cdot \frac{c}{d} = \frac{ac}{bd}.$$

$$\begin{aligned}
\text{For} \quad \frac{a}{b} \cdot \frac{c}{d} \cdot bd &= \frac{a}{b} b \cdot \frac{c}{d} d, && \text{Laws IV, III.}\\
&= ac, && \text{Def. VIII.}\\
\text{and} \quad \frac{ac}{bd} \cdot bd &= ac. && \text{Def VIII.}
\end{aligned}$$

The theorem follows by law IX.

2.
$$\frac{\frac{a}{b}}{\frac{c}{d}} = \frac{ad}{bc},$$

$$\begin{aligned}
\text{For} \quad \frac{\frac{a}{b}}{\frac{c}{d}} \cdot \frac{c}{d} &= \frac{a}{b}, && \text{Def. VIII.}\\
\text{and} \quad \frac{ad}{bc} \cdot \frac{c}{d} &= \frac{a}{b} \cdot \frac{dc}{cd}, && \text{§18, 1; Law IV.}\\
&= \frac{a}{b}, &&\\
\text{since} \quad \frac{dc}{cd} &= dc = 1 \times cd. && \text{Def. VIII, Law IX.}
\end{aligned}$$

The theorem follows by law IX.

3.
$$\frac{a}{b} \pm \frac{c}{d} = \frac{ad \pm bc}{bd}.$$

$$\begin{aligned}
\text{For} \quad \left(\frac{a}{b} \pm \frac{c}{d}\right) bd &= \frac{a}{b} b \cdot d \pm \frac{c}{d} d \cdot b, && \text{Laws III–V: §10, 5.}\\
&= ad \pm bc, && \text{Def. VIII.}\\
\text{and} \quad \left(\frac{ad \pm bc}{bd}\right) bd &= ad \pm bc. && \text{Def. VIII.}
\end{aligned}$$

The theorem follows by law IX.

By the same method it may be inferred that

4.
$$\begin{aligned}
\frac{a}{b} &>, \quad =, < \frac{c}{d},\\
\text{as} \quad ad &>, \quad =, < bc. \quad \text{Def. VIII, Laws III, IV, IX, IX'.}
\end{aligned}$$

19. Limitations of Numerical Division. Symbolic Division. The Fraction. General as is the form of the preceding equations, they are capable of numerical interpretation only when $\frac{a}{b}$, $\frac{c}{d}$ are numbers, a case of comparatively rare occurrence. The narrow limits set the quotient in the numerical definition render division an unimportant operation as compared with addition, multiplication, or the generalized subtraction discussed in the preceding chapter.

But the way which led to an unrestricted subtraction lies open also to the removal of this restriction; and the reasons for following it there are even more cogent here.

We accept as the quotient of a divided by any number b, which is not 0, the symbol $\frac{a}{b}$ defined by the equation

$$\left(\frac{a}{b}\right) b = a,$$

regarding this equation merely as a declaration of the equivalence of the symbols $(\frac{a}{b})b$ and a, of the right to substitute one for the other in any reckoning.

Whether $\frac{a}{b}$ be a number or not is to this definition irrelevant. When a mere symbol, $\frac{a}{b}$ is called a *fraction*, and in opposition to this a number is called an *integer*.

We then put ourselves in immediate possession of definitions of the addition, subtraction, multiplication, and division of this symbol, as well as of the relations of equality and greater and lesser inequality—definitions which are consistent with the corresponding numerical definitions and with one another—by assuming the permanence of form of the equations 1, 2, 3 and of the test 4 of § 18 as symbolic statements, when they cease to be interpretable as numerical statements.

The purely symbolic character of $\frac{a}{b}$ and its operations detracts nothing from their legitimacy, and they establish division on a footing of at least formal equality with the other three fundamental operations of arithmetic.[2]

20. Negative Fractions. Inasmuch as negatives conform to the laws and definitions I–IX, the equations 1, 2, 3 and the test 4 of §18 are valid when any of the numbers a, b, c, d are replaced by negatives. In particular, it follows from the definition of quotient and its determinateness, that

$$\frac{a}{-b} = -\frac{a}{b}; \frac{-a}{b} = -\frac{a}{b}; \frac{-a}{-b} = \frac{a}{b}.$$

It ought, perhaps, to be said that the determinateness of division of negatives has not been formally demonstrated. The theorem, however, that if $(\pm a)(\pm c) = (\pm b)(\pm c), \pm a = \pm b$, follows for every selection of the signs $\pm$ from the one selection $+, +, +, +$ by §14, 6, 8.

21. General Test of the Equality or Inequality of Fractions.

Given any two fractions $\pm\frac{a}{b}, \pm\frac{c}{d}$.

$$\pm\tfrac{a}{b} >, \quad = \text{ or } < \pm\tfrac{c}{d},$$

$$\text{according as } \pm ad >, \quad = \text{ or } < \pm bc.$$

Laws IX, IX'. Compare §4, §14, 9.

22. Indeterminateness of Division by Zero. Division by 0 does not conform to the law of determinateness; the equations 1, 2, 3 and the test 4 of § 18 are, therefore, not valid when 0 is one of the divisors.

The symbols $\frac{0}{0}$, $\frac{a}{0}$, of which some use is made in mathematics, are indeterminate.[3]

---

[2]The doctrine of symbolic division admits of being presented in the very same form as that of symbolic subtraction.

The equations of Chapter II immediately pass over into theorems respecting division when the signs of multiplication and division are substituted for those of addition and subtraction; so, for instance,

$$a - (b + c) = a - b - c = a - c - b \text{ gives } \frac{a}{bc} = \frac{(\frac{a}{b})}{c} = \frac{(\frac{a}{c})}{b}$$

In particular, to $(a - a) + a = a$ corresponds $\frac{a}{a}a = a$. Thus a purely symbolic definition may be given 1. It plays the same rôle in multiplication as 0 in addition. Again, it has the same exceptional character in involution—an operation related to multiplication quite as multiplication to addition—as 0 in multiplication; for $1^m = 1^n$, whatever the values of m and n.

Similarly, to the equation $(-a) + a = 0$, or $(0 - a) + a = 0$, corresponds $(\frac{1}{a})a = 1$, which answers as a definition of the unit fraction $\frac{1}{a}$; and in terms of these unit fractions and integers all other fractions may be expressed.

[3]In this connection see § 32.

1. $\frac{0}{0}$ is indeterminate. For $\frac{0}{0}$ is completely defined by the equation $\left(\frac{0}{0}\right)0 = 0$; but since $x \times 0 = 0$, whatever the value of x, any number whatsoever will satisfy this equation.

2. $\frac{a}{0}$ is indeterminate. For, by definition, $\left(\frac{a}{0}\right)0 = a$. Were $\frac{a}{0}$ determinate, therefore,—since then $\left(\frac{a}{0}\right)0$ would, by § 18, 1, be equal to $\frac{a \times 0}{0}$, or to $\frac{0}{0}$,—the number a would be equal to $\frac{0}{0}$, or indeterminate.

Division by 0 is not an admissible operation.

23. Determinateness of Symbolic Division. This exception to the determinateness of division may seem to raise an objection to the legitimacy of assuming—as is done when the demonstrations 1–4 of § 18 are made to apply to symbolic quotients—that symbolic division is determinate.

It must be observed, however, that $\frac{0}{0}$, $\frac{a}{0}$ are indeterminate in the *numerical* sense, whereas by the determinateness of symbolic division is, of course, not meant actual numerical determinateness, but "symbolic determinateness," conformity to law IX, taken merely as a symbolic statement. For, as has been already frequently said, from the present standpoint the *fraction* $\frac{a}{b}$ is a mere symbol, altogether without numerical meaning apart from the equation $\left(\frac{a}{b}\right)b = a$, with which, therefore, the property of numerical determinateness has no possible connection. The same is true of the product, sum or difference of two fractions, and of the quotient of one fraction by another.

As for symbolic determinateness, it needs no justification when assumed, as in the case of the fraction and the demonstrations 1–4, of symbols whose definitions do not preclude it. The inference, for instance, that because

$$\left(\frac{a}{b}\frac{c}{d}\right)bd = \left(\frac{ac}{bd}\right)bd,$$
$$\frac{a}{b}\frac{c}{d} = \frac{ac}{bd},$$

which depends on this principle of symbolic determinateness, is of precisely the same character as the inference that

$$\left(\frac{a}{b}\frac{c}{d}\right) = \frac{a}{b}b \cdot \frac{c}{d}d,$$

which depends on the associative and commutative laws.

Both are pure assumptions made of the *undefined* symbol $\frac{a}{b}\frac{c}{d}$ for the sake of securing it a definition identical in form with that of the product of two numerical quotients.[4]

24. The Vanishing of a Product. It has already been shown (§ 13, 3, § 14, 7, § 18, 1) that the sufficient condition for the vanishing of a product is the vanishing of one of its factors. From the determinateness of division it follows that this is also the necessary condition, that is to say:

If a product vanish, one of its factors must vanish.

Let $xy = 0$, where x, y may represent numbers or any of the symbols we have been considering.

---

[4]These remarks, *mutatis mutandis*, apply with equal force to subtraction.

Since $$xy = 0,$$

$$xy + xz = xz, \qquad \S 13, 1.$$

or $$x(y + z) = xz, \qquad \text{Law V.}$$

whence, if x be not 0, $$y + z = z, \qquad \text{Law IX.}$$

or $$y = 0. \qquad \text{Law VII.}$$

25. The System of Rational Numbers. Three symbols, 0, $-d$, $\frac{a}{b}$, have thus been found which can be reckoned with by the same rules as numbers, and in terms of which it is possible to express the result of every addition, subtraction, multiplication or division, whether performed on numbers or on these symbols themselves; therefore, also, the result of any complex operation which can be resolved into a finite combination of these four operations.

Inasmuch as these symbols play the same rôle as numbers in relation to the fundamental operations of arithmetic, it is natural to class them with numbers. The word "number," originally applicable to the positive integer only, has come to apply to zero, the negative integer, the positive and negative fraction also, this entire group of symbols being called the system of *rational numbers*.[5] This involves, of course, a radical change of the number concept, in consequence of which numbers become merely part of the symbolic equipment of certain operations, admitting, for the most part, of only such definitions as these operations lend them.

In accepting these symbols as its numbers, arithmetic ceases to be occupied exclusively or even principally with the properties of numbers in the strict sense. It becomes an *algebra*, whose immediate concern is with certain operations defined, as addition by the equations $a + b = b + a$, $a + (b + c) = a + b + c$, formally only, without reference to the meaning of the symbols operated on.[6]

[5]It hardly need be said that the fraction, zero, and the negative actually made their way into the number-system for quite a different reason from this;—because they admitted of certain "real" interpretations, the fraction in measurements of lines, the negative in debit where the corresponding positive meant credit or in a length measured to the left where the corresponding positive meant a length measured to the right. Such interpretations, or correspondences to existing things which lie entirely outside of pure arithmetic, are ignored in the present discussion as being irrelevant to a pure arithmetical doctrine of the artificial forms of number.

[6]The word "algebra" is here used in the general sense, the sense in which *quaternions* and the *Ausdehungslehre* (see §§ 127, 128) are algebras. Inasmuch as elementary arithmetic, as actually constituted, accepts the fraction, there is no essential difference between it and elementary algebra with respect to the kinds of number with which it deals; algebra merely goes further in the use of artificial numbers. The elementary algebra differs from arithmetic in employing literal symbols for numbers, but chiefly in making the equation an object of investigation.

4. THE IRRATIONAL.

26. The System of Rational Numbers Inadequate. The system of rational numbers, while it suffices for the four fundamental operations of arithmetic and finite combinations of these operations, does not fully meet the needs of algebra.

The great central problem of algebra is the equation, and that only is an adequate number-system for algebra which supplies the means of expressing the roots of all possible equations. The system of rational numbers, however, is equal to the requirements of equations of the first degree only; it contains symbols not even for the roots of such elementary equations of higher degrees as $x^2 = 2$, $x^2 = -1$.

But how is the system of rational numbers to be enlarged into an algebraic system which shall be adequate and at the same time sufficiently simple?

The roots of the equation

$$x^n + p_1 x^{n-1} + p_2 x^{n-2} + \cdots + p_{n-1} x + p_n = 0$$

are not the results of single elementary operations, as are the negative of subtraction and the fraction of division; for though the roots of the quadratic are results of "evolution," and the same operation often enough repeated yields the roots of the cubic and biquadratic also, it fails to yield the roots of higher equations. A system built up as the rational system was built, by accepting indiscriminately every new symbol which could show cause for recognition, would, therefore, fall in pieces of its own weight.

The most general characteristics of the roots must be discovered and defined and embodied in symbols—by a method which does not depend on processes for solving equations. These symbols, of course, however characterized otherwise, must stand in consistent relations with the system of rational numbers and their operations.

An investigation shows that the forms of number necessary to complete the algebraic system may be reduced to two: the symbol $\sqrt{-1}$, called the *imaginary* (an indicated root of the equation $x^2 + 1 = 0$), and the class of symbols called *irrational*, to which the roots of the equation $x^2 - 2 = 0$ belong.

27. Numbers Defined by Regular Sequences. The Irrational. On applying to 2 the ordinary method for extracting the square root of a number, there is obtained the following sequence of numbers, the results of carrying the reckoning out to 0, 1, 2, 3, 4, ... places of decimals, viz.:

$$1, 1.4, 1.41, 1.414, 1.4142, \ \ldots$$

These numbers are rational; the first of them differs from each that follows it by less than 1, the second by less than $\frac{1}{10}$, the third by less than $\frac{1}{100}$, ... the nth by less than $\frac{1}{10^{n-1}}$. And $\frac{1}{10^{n-1}}$ is a fraction which may be made less than any assignable number whatsoever by taking n great enough.

This sequence may be regarded as a definition of the square root of 2. It is such in the sense that a term may be found in it the square of which, as well as of each following term, differs from 2 by less than any assignable number.

Any sequence of rational numbers

$$\alpha_1, \alpha_2, \alpha_3, \cdots, \alpha_\mu, \alpha_{\mu+1}, \cdots \alpha_{\mu+\nu}, \cdots$$

in which, as in the above sequence, the term α_μ may, by taking μ great enough, be made to differ numerically from each term that follows it by less than any assignable *number, so that, for all values of* ν, *the difference,* $\alpha_{\mu+\nu} - \alpha_\mu$, *is numerically less than* δ, *however small* δ *be taken, is called a regular sequence.*

The entire class of operations which lead to regular sequences may be called *regular sequence-building.* Evolution is only one of many operations belonging to this class.

Any regular sequence is said to "define a number,"—this "number" being merely the symbolic, ideal, result of the operation which led to the sequence. It will sometimes be convenient to represent numbers thus defined by the single letters a, b, c, etc., which have heretofore represented positive integers only.

After some particular term all terms of the sequence $\alpha_1, \alpha_2, \cdots$ may be the same, say α. The number defined by the sequence is then α itself. A place is thus provided for rational numbers in the general scheme of numbers which the definition contemplates.

When not a rational, the number defined by a regular sequence is called *irrational.*

The regular sequence .3, .33, . . . , has a *limiting value,* viz., $\frac{1}{3}$; which is to say that a term can be found in this sequence which itself, as well as each term which follows it, differs from $\frac{1}{3}$ by less than any assignable number. In other words, the difference between $\frac{1}{3}$ and the μth term of the sequence may be made less than any assignable number whatsoever by taking μ great enough. It will be shown presently that the number defined by any regular sequence, $\alpha_1, \alpha_2, \cdots$ stands in this same relation to its term α_μ.

28. Zero, Positive, Negative. In any regular sequence $\alpha_1, \alpha_2, \cdots$ a term α_μ may always be found which itself, as well as each term which follows it, is either

(1) numerically less than any assignable number,
or (2) greater than some definite positive rational number,
or (3) less than some definite negative rational number.

In the first case the number a, which the sequence defines, is said to be *zero,* in the second *positive,* in the third *negative.*

29. The Four Fundamental Operations. *Of the numbers defined by the two sequences:*

$$\alpha_1, \alpha_2, \alpha_3, \cdots, \alpha_\mu, \alpha_{\mu+1}, \cdots, \alpha_{\mu+\nu}, \cdots,$$
$$\beta_1, \beta_2, \beta_3, \cdots, \beta_\mu, \beta_{\mu+1}, \cdots, \beta_{\mu+\nu}, \cdots$$

(1) *The sum is the number defined by the sequence:*

$$\alpha_1 + \beta_1, \alpha_2 + \beta_2, \cdots \alpha_\mu + \beta_\mu, \alpha_{\mu+1} + \beta_{\mu+1}, \cdots \alpha_{\mu+\nu} + \beta_{\mu+\nu}, \cdots$$

(2) *The difference is the number defined by the sequence:*

$$\alpha_1 - \beta_1, \alpha_2 - \beta_2, \cdots \alpha_\mu - \beta_\mu, \alpha_{\mu+1} - \beta_{\mu+1}, \cdots \alpha_{\mu+\nu} - \beta_{\mu+\nu}, \cdots$$

(3) *The product is the number defined by the sequence:*

$$\alpha_1\beta_1, \alpha_2\beta_2, \cdots \alpha_\mu\beta_\mu, \alpha_{\mu+1}\beta_{\mu+1}, \cdots \alpha_{\mu+\nu}\beta_{\mu+\nu}, \cdots$$

(4) *The quotient is the number defined by the sequence:*

$$\frac{\alpha_1}{\beta_1}, \frac{\alpha_2}{\beta_2}, \cdots \frac{\alpha_\mu}{\beta_\mu}, \frac{\alpha_{\mu+1}}{\beta_{\mu+1}}, \cdots \frac{\alpha_{\mu+\nu}}{\beta_{\mu+\nu}}, \cdots$$

For these definitions are consistent with the corresponding definitions for rational numbers; they reduce to these elementary definitions, in fact, whenever the sequences

$\alpha_1, \alpha_2, \ldots; \beta_1, \beta_2, \ldots$ either reduce to the forms $\alpha, \alpha, \ldots; \beta, \beta, \ldots$ or have rational limiting values.

They conform to the fundamental laws I–IX. This is immediately obvious with respect to the commutative, associative, and distributive laws, the corresponding terms of the two sequences $\alpha_1\beta_1, \alpha_2\beta_2, \ldots; \beta_1\alpha_1, \beta_2\alpha_2, \ldots$, for instance, being identically equal, by the commutative law for rationals.

But again division as just defined is determinate. For division can be indeterminate only when a product may vanish without either factor vanishing (cf. § 24); whereas $\alpha_1\beta_1, \alpha_2\beta_2, \ldots$ can define 0, or its terms after the nth fall below any assignable number whatsoever, only when the same is true of one of the sequences $\alpha_1, \alpha_2, \ldots; \beta_1, \beta_2, \ldots$ [1]

It only remains to prove, therefore, that the sequences (1), (2), (3), (4) are qualified to define numbers (§ 27).

(1) and (2) Since the sequences $\alpha_1, \alpha_2, \ldots; \beta_1, \beta_2, \ldots$ are, by hypothesis, such as define numbers, corresponding terms in the two, α_μ, β_μ may be found, such that

$$\alpha_{\mu+\nu} - \alpha_\mu \quad \text{is numerically} \quad < \delta,$$

and $$\beta_{\mu+\nu} - \beta_\mu \quad \text{is numerically} \quad < \delta,$$

and, therefore, $$(\alpha_{\mu+\nu} \pm \beta_{\mu+\nu}) - (\alpha_\mu \pm \beta_\mu) < 2\delta,$$

for all values of ν, and that however small δ may be.

Therefore each of the sequences $\alpha_1+\beta_1, \alpha_2+\beta_2, \ldots; \alpha_1-\beta_1, \alpha_2-\beta_2, \ldots$ is regular.

(3) Let α_μ and β_μ be chosen as before.

Then $\alpha_{\mu+\nu}\beta_{\mu+\nu} - \alpha_\mu\beta_\mu$,

since it is identically equal to

$$\alpha_{\mu+\nu}(\beta_{\mu+\nu} - \beta_\mu) + \beta_\mu(\alpha_{\mu+\nu} - \alpha_\mu),$$

is numerically less than $\alpha_{\mu+\nu}\delta + \beta_\mu\delta$, and may, therefore, be made less than any assignable number by taking δ small enough; and that for all values of ν.

Therefore the sequence $\alpha_1\beta_1, \alpha_2\beta_2, \ldots$ is regular.

$$(4) \qquad \frac{\alpha_{\mu+\nu}}{\beta_{\mu+\nu}} - \frac{\alpha_\mu}{\beta_\mu} = \frac{\alpha_{\mu+\nu}\beta_\mu - \beta_{\mu+\nu}\alpha_\mu}{\beta_{\mu+\nu}\beta_\mu},$$

which is identically equal to

$$\frac{\beta_{\mu+\nu}(\alpha_{\mu+\nu} - \alpha_\mu) - \alpha_{\mu+\nu}(\beta_{\mu+\nu} - \beta_\mu)}{\beta_{\mu+\nu}\beta_\mu}.$$

By choosing α_μ and β_μ as before the numerator of this fraction, and therefore the fraction itself, may be made less than any assignable number; and that for all values of ν.

Therefore the sequence $\frac{\alpha_1}{\beta_1}, \frac{\alpha_2}{\beta_2}, \ldots$ is regular.

30. Equality. Greater and Lesser Inequality. *Of two numbers, a and b, defined by regular sequences* $\alpha_1, \alpha_2, \ldots,; \beta_1, \beta_2, \ldots$, *the first is greater than, equal to or less than the second, according as the number defined by* $\alpha_1 - \beta_1, \alpha_1 - \beta_2, \ldots$ *is greater than, equal to or less than* 0.

This definition is to be justified exactly as the definitions of the fundamental operations on numbers defined by regular sequences were justified in § 29.

From this definition, and the definition of 0 in § 28, it immediately follows that

[1] It is worth noticing that the determinateness of division is here not an independent assumption, but a consequence of the definition of multiplication and the determinateness of the division of rationals. The same thing is true of the other fundamental laws I–V, VII.

COR. *Two numbers which differ by less than any assignable number are equal.*

31. The Number Defined by a Regular Sequence is its Limiting Value. The difference between a number a and the term α_μ of the sequence by which it is defined may be made less than any assignable number by taking μ great enough.

For it is only a restatement of the definition of a regular sequence $\alpha_1, \alpha_2, \ldots$ to say that the sequence

$$\alpha_1 - \alpha_\mu, \alpha_2 - \alpha_\mu, \ldots, \alpha_{\mu+\nu} - \alpha_\mu, \ldots,$$

which defines the difference $a - \alpha_\mu$ (§ 29, 2), is one whose terms after the μth can be made less than any assignable number by choosing μ great enough, and which, therefore, becomes, as μ is indefinitely increased, a sequence which defines 0 (§ 28).

In other words, the *limit* of $a - \alpha_\mu$ as μ is indefinitely increased is 0, or $a =$ limit (α_μ). Hence

The number defined by a regular sequence is the limit to which the μth term of this sequence approaches as μ is indefinitely increased.[2]

The definitions (1), (2), (3), (4) of § 29 may, therefore, be stated in the form:

$$\begin{aligned} \text{limit}\,(\alpha_\mu) \pm \text{limit}\,(\beta_\mu) &= \text{limit}\,(\alpha_\mu \pm \beta_\mu), \\ \text{limit}\,(\alpha_\mu) \cdot \text{limit}\,(\beta_\mu) &= \text{limit}\,(\alpha_\mu \beta_\mu), \\ \frac{\text{limit}\,(\alpha_\mu)}{\text{limit}\,(\beta_\mu)} &= \text{limit}\left(\frac{\alpha_\mu}{\beta_\mu}\right). \end{aligned}$$

For limit (α_μ) the more complete symbol $\lim\limits_{\mu \doteq \infty} (\alpha_\mu)$ is also used, read "the limit which α_μ approaches as μ approaches infinity"; the phrase "approaches infinity" meaning only, "becomes greater than any assignable number."

32. Division by Zero. (1) The sequence $\dfrac{\alpha_1}{\beta_1}, \dfrac{\alpha_2}{\beta_2}, \ldots$ cannot define a number when the number defined by $\beta_1, \beta_2, \ldots$ is 0, unless the number defined by $\alpha_1, \alpha_2, \ldots$ be also 0. In this case it may; $\dfrac{\alpha_\mu}{\beta_\mu}$ may approach a definite limit as μ increases, however small α_μ and β_μ become. But this number is not to be regarded as the mere quotient $\dfrac{0}{0}$. Its value is not at all determined by the fact that the numbers defined by $\alpha_1, \alpha_2, \ldots$; $\beta_1, \beta_2, \ldots$ are 0; for there is an indefinite number of different sequences which define 0, and by properly choosing $\alpha_1, \alpha_2, \ldots$; $\beta_1, \beta_2, \ldots$ from among them, the terms of the sequence $\dfrac{\alpha_1}{\beta_1}, \dfrac{\alpha_2}{\beta_2}, \ldots$ may be made to take any value whatsoever.

(2) The sequence $\dfrac{\alpha_1}{\beta_1}, \dfrac{\alpha_2}{\beta_2}, \ldots$ is not regular when $\beta_1, \beta_2, \ldots$ defines 0 and $\alpha_1, \alpha_2, \ldots$ defines a number different from 0.

No term $\dfrac{\alpha_\mu}{\beta_\mu}$ can be found which differs from the terms following it by less than any assignable number; but rather, by taking μ great enough, $\dfrac{\alpha_\mu}{\beta_\mu}$ can be made greater

---

[2]What the above demonstration proves is that a stands in the same relation to α_μ when irrational as when rational. The principle of permanence (cf. § 12), therefore, justifies one in regarding a as the ideal limit in the former case since it is the actual limit in the latter (§ 27). a, when irrational, is limit (α_μ) in precisely the same sense that $\dfrac{c}{d}$ is the quotient of c by d, when c is a positive integer not containing d. It follows from the demonstration that if there be a reality corresponding to a, as in geometry we assume there is (§ 40), that reality will be the actual limit of the reality of the same kind corresponding to α_μ.

The notion of irrational limiting values was not immediately available because, prior to §§ 28, 29, 30, the meaning of difference and greater and lesser inequality had not been determined for numbers defined by sequences.

than any assignable number whatsoever.

Though not regular and though they do not define numbers, such sequences are found useful in the higher mathematics. They may be said to define *infinity.* Their usefulness is due to their determinate form, which makes it possible to bring them into combination with other sequences of like character or even with regular sequences.

Thus the quotient of any regular sequence $\gamma_1, \gamma_2, \ldots$ by $\frac{\alpha_1}{\beta_1}, \frac{\alpha_2}{\beta_2}, \ldots$ is a regular sequence and defines 0; and the quotient of $\frac{\alpha_1}{\beta_1}, \frac{\alpha_2}{\beta_2}, \ldots$ by a similar sequence $\frac{\gamma_1}{\delta_1}, \frac{\gamma_2}{\delta_2}, \ldots$ may also be regular and serve—if α_i, β_i, γ_i, δ_i $(i = 1, 2, \ldots)$ be properly chosen—to define any number whatsoever.

The term $\frac{\alpha_\mu}{\beta_\mu}$ "approaches infinity" (*i. e.* increases without limit) as μ is indefinitely increased, in a definite or determinate manner; so that the infinity which $\frac{\alpha_1}{\beta_1}, \frac{\alpha_2}{\beta_2}, \ldots$ defines is not indeterminate like the mere symbol $\frac{a}{0}$ of § 22.

But here again it is to be said that this determinateness is not due to the mere fact that $\beta_1, \beta_2 \ldots$ defines 0, which is all that the unqualified symbol $\frac{a}{0}$ expresses. For there is an indefinite number of different sequences which like $\beta_1, \beta_2, \ldots$ define 0, and $\frac{a}{0}$ is a symbol for the quotient of a by any one of them.

33. The System defined by Regular Sequences of Rationals, Closed and Continuous. *A regular sequence of irrationals*

$$a_1, a_2, \ldots a_m, a_{m+1}, \ldots a_{m+n}, \ldots$$

(in which the differences $a_{m+n} - a_m$ may be made numerically less than any assignable number by taking m great enough) defines a number, but never a number which may not also be defined by a sequence of rational numbers.

For $\beta_1, \beta_2, \ldots$ being any sequence of rationals which defines 0, construct a sequence of rationals $\alpha_1, \alpha_2, \ldots$ such that $a_1 - \alpha_1$ is numerically less than β_1 (§ 30), and in the same sense $a_2 - \alpha_2 < \beta_2$, $a_3 - \alpha_3 < \beta_3$ etc. Then limit $(a_m - \alpha_m) = 0$ (§§ 28, 31), or limit $(a_m) = \text{limit}(\alpha_m)$.

This theorem justifies the use of regular sequences of irrationals for defining numbers, and so makes possible a simple expression of the results of some very complex operations. Thus a^m, where m is irrational, is a number; the number, namely, which the sequence $a^{\alpha_1}, a^{\alpha_2}, \ldots$ defines, when $\alpha_1, \alpha_2, \ldots$ is any sequence of rationals defining m.

But the importance of the theorem in the present discussion lies in its declaration that the number-system defined by regular sequences of rationals contains all numbers which result from the operations of regular sequence-building in general. It is a *closed* system with respect to the four fundamental operations and this new operation, exactly as the rational numbers constitute a closed system with respect to the four fundamental operations only (cf. § 25).

The system of numbers defined by regular sequences of rationals—*real* numbers, as they are called—therefore possesses the following two properties: (1) between every two unequal, real numbers there are other real numbers; (2) a variable which runs through any regular sequence of real numbers, rational or irrational, will approach a real number as limit. We indicate all this by saying that the system of real numbers is **continuous**.

5. THE IMAGINARY. COMPLEX NUMBERS.

34. The Pure Imaginary. The other symbol which is needed to complete the number-system of algebra, unlike the irrational but like the negative and the fraction, admits of definition by a single equation of a very simple form, viz.,

$$x^2 + 1 = 0$$

It is the symbol whose square is -1, the symbol $\sqrt{-1}$, now commonly written i.[1] It is called the *unit of imaginaries.*

In contradistinction to i all the forms of number hitherto considered are called *real.* These names, "real" and "imaginary," are unfortunate, for they suggest an opposition which does not exist. Judged by the only standards which are admissible in a pure doctrine of numbers i is imaginary in the same sense as the negative, the fraction, and the irrational, but in no other sense; all are alike mere symbols devised for the sake of representing the results of operations even when these results are not numbers (positive integers). i got the name imaginary from the difficulty once found in discovering some extra-arithmetical reality to correspond to it.

As the only property attached to i by definition is that its square is -1, nothing stands in the way of its being "multiplied" by any real number a; the product, ia, is called a *pure imaginary.*

An entire new system of numbers is thus created, coextensive with the system of real numbers, but distinct from it. Except 0, there is no number in the one which is at the same time contained in the other.[2] Numbers in either system may be compared with each other by the definitions of equality and greater and lesser inequality (§ 30), ia being called $\gtreqless ib$, as $a \gtreqless b$; but a number in one system cannot be said to be either greater than, equal to or less than a number in the other system.

35. Complex Numbers. The sum $a + ib$ is called a *complex number.* Its terms belong to two distinct systems, of which the fundamental units are 1 and i.

The *general* complex number $a + ib$ is defined by a *complex sequence*

$$\alpha_1 + i\beta_1,\ \alpha_2 + i\beta_2, \ldots, \alpha_\mu + i\beta_\mu, \ldots,$$

where $\alpha_1, \alpha_2, \ldots;\ \beta_1, \beta_2, \ldots$ are regular sequences.

Since $a = a + i0$ (§ 36, 3, Cor.) and $ib = 0 + ib$, all real numbers, a, and pure imaginaries, ib, are contained in the system of complex numbers $a + ib$.

$a + ib$ can vanish only when both $a = 0$ and $b = 0$.

36. The Four Fundamental Operations on Complex Numbers. The assumption of the permanence of the fundamental laws leads immediately to the following definitions of the addition, subtraction, multiplication, and division of complex numbers.

---

[1] Gauss introduced the use of i to represent $\sqrt{-1}$.

[2] Throughout this discussion ∞ is not regarded as belonging to the number-system, but as a limit of the system, lying without it, a symbol for something greater than any number of the system.

1. $(a+ib)+(a'+ib') = a+a'+i(b+b')$.

For $(a+ib)+(a'+ib') = a+ib+a'+ib'$, Law II.

$= a+a'+ib+ib'$, Law I.

$= a+a'+i(b+b')$. Laws II, V.

2. $(a+ib)-(a'+ib') = a-a'+i(b-b')$.

By definition of subtraction (VI) and § 36, 1.

COR. *The necessary as well as the sufficient condition for the equality of two complex numbers $a+ib$, $a'+ib'$ is that $a=a'$ and $b=b'$.*

$$\text{For if}\quad (a+ib)-(a'+ib') = a-a'+i(b-b') = 0,$$
$$a-a'=0, b-b'=0\ (\S\ 35),\ \text{or}\ a=a', b=b'.$$

3. $(a+ib)(a'+ib') = aa'-bb'+i(ab'+ba')$.

For $(a+ib)(a'+ib') = (a+ib)a'+(a+ib)ib'$, Law V.

$= aa'+ib\cdot a'+a\cdot ib'+ib\cdot ib'$, Law V.

$= (aa'-bb')+i(ab'+ba')$. Laws I–V.

COR. *If either factor of a product vanish, the product vanishes.*

$$\text{For}\quad i\times 0 = i(b-b) = ib-ib\ (\S\ 10,5), = 0\ (\S\ 14,1).$$
$$\text{Hence}\quad (a+ib)0 = a\times 0+ib\times 0 = a\times 0+i(b\times 0) = 0.$$

Laws V, IV, § 28, § 29, 3.

4. $$\frac{a+ib}{a'+ib'} = \frac{aa'+bb'}{a'^2+b'^2}+i\frac{ba'-ab'}{a'^2+b'^2}.$$

For let the quotient of $a+ib$ by $a'+ib'$ be $x+iy$.

By the definition of division (VIII),

$$(x+iy)(a'+ib') = a+ib.$$
$$\therefore\quad xa'-yb'+i(xb'+ya') = a+ib. \qquad \S\ 36,3$$
$$\therefore\quad xa'-yb' = a,\ xb'+ya' = b. \qquad \S\ 36,2, Cor.$$

Hence, solving for x and y between these two equations,

$$x = \frac{aa'+bb'}{a'^2+b'^2},\quad y = \frac{ba'-ab'}{a'^2+b'^2}.$$

Therefore, as in the case of real numbers, division is a determinate operation, except when the divisor is 0; it is then indeterminate. For x and y are determinate (by IX) unless $a'^2+b'^2=0$, that is, unless $a'=b'=0$, or $a'+ib'=0$; for a' and b' being real, a'^2 and b'^2 are both positive, and one cannot destroy the other.[3] Hence, by the reasoning in § 24,

[3]What is here proven is that in the system of complex numbers formed from the fundamental units 1 and i there is one, and but one, number which is the quotient of $a+ib$ by $a'+ib'$; this being a consequence of the determinateness of the division of real numbers and the peculiar relation ($i^2=-1$) holding between the fundamental units. For the sake of the permanence of IX we make the assumption, otherwise irrelevant, that this is the only value of the quotient whether within or without the system formed from the units 1 and i.

COR. *If a product of two complex numbers vanish, one of the factors must vanish.*

37. Numerical Comparison of Complex Numbers. Two complex numbers, $a+ib$, $a'+ib'$, do not, generally speaking, admit of direct comparison with each other, as do two real numbers or two pure imaginaries; for a may be greater than a', while b is less than b'.

They are compared *numerically*, however, by means of their *moduli* $\sqrt{a^2+b^2}$, $\sqrt{a'^2+b'^2}$; $a+ib$ being said to be numerically greater than, equal to or less than $a'+ib'$ according as $\sqrt{a^2+b^2}$ is greater than, equal to or less than $\sqrt{a'^2+b'^2}$. Compare § 47.

38. The Complex System Adequate. The system $a+ib$ is an adequate number-system for algebra. For, as will be shown (Chapter VII), all roots of algebraic equations are contained in this system.

But more than this, the system $a+ib$ is a closed system with respect to all existing mathematical operations, as are the rational system with respect to all finite combinations of the four fundamental operations and the real system with respect to these operations and regular sequence-building. For the results of the four fundamental operations on complex numbers are complex numbers (§ 36, 1, 2, 3, 4). Any other operation may be resolved into either a finite combination of additions, subtractions, multiplications, divisions or such combinations indefinitely repeated. In either case the result, if determinate, is a complex number, as follows from the definitions 1, 2, 3, 4 of § 36, and the nature of the real number-system as developed in the preceding chapter (see Chapter VIII).

The most important class of these higher operations, and the class to which the rest may be reduced, consists of those operations which result in infinite series (Chapter VIII); among which are involution, evolution, and the taking of logarithms (Chapter IX), sometimes included among the fundamental operations of algebra.

39. Fundamental Characteristics of the Algebra of Number. The algebra of number is completely characterized, formally considered, by the laws and definitions I–IX and the fact that its numbers are expressible linearly in terms of two fundamental units.[4] It is a linear, associative, distributive, commutative algebra. Moreover, the most general linear, associative, distributive, commutative algebra, whose numbers are complex numbers of the form $x_1e_1+x_2e_2+\cdots+x_ne_n$, built from n fundamental units $e_1, e_2, \ldots, e_n$, is reducible to the algebra of the complex number $a+ib$. For Weierstrass[5] has shown that any two complex numbers a and b of the form $x_1e_1+x_2e_2+\cdots+x_ne_n$, whose sum, difference, product, and quotient are numbers of this same form, and for which the laws and definitions I–IX hold good, may by suitable transformations be resolved into components $a_1, a_2, \ldots a_r$; $b_1, b_2, \ldots b_r$, such that

[4]That is, in terms of the first powers of these units.

[5]Zur Theorie der aus n Haupteinheiten gebildeten complexen Grössen. Göttinger Nachrichten Nr. 10, 1884.

Weierstrass finds that these general complex numbers differ in only one important respect from the complex number $a+ib$. If the number of fundamental units be greater than 2, there always exist numbers, different from 0, the product of which by certain other numbers is 0. Weierstrass calls them divisors of 0. The number of exceptions to the determinateness of division is infinite instead of one.

$$\begin{aligned}a &= a_1 + a_2 + \cdots + a_r,\\ b &= b_1 + b_2 + \cdots + b_r,\\ a \pm b &= a_1 \pm b_1 + a_2 \pm b_2 + \cdots + a_r \pm b_r,\\ ab &= a_1 b_1 + a_2 b_2 + \cdots + a_r b_r,\\ \frac{a}{b} &= \frac{a_1}{b_1} + \frac{a_2}{b_2} + \cdots + \frac{a_r}{b_r}.\end{aligned}$$

The components a_i, b_i are constructed either from one fundamental unit g_i or from two fundamental units g_i, k_i.[6]

For components of the first kind the multiplication formula is

$$(\alpha g_i)(\beta g_i) = (\alpha\beta) g_i.$$

For components of the second kind the multiplication formula is

$$(\alpha g_i + \beta k_i)(\alpha' g_i + \beta' k_i) = (\alpha\alpha' - \beta\beta') g_i + (\alpha\beta' + \beta\alpha') k_i.$$

And these formulas are evidently identical with the multiplication formulas

$$\begin{aligned}(\alpha 1)(\beta 1) &= (\alpha\beta) 1,\\ (\alpha 1 + \beta i)(\alpha' 1 + \beta' i) &= (\alpha\alpha' - \beta\beta') 1 + (\alpha\beta' + \beta\alpha') i\end{aligned}$$

of common algebra.

[6]These units are, generally speaking, not $e_1, e_2, \ldots, e_n$, but linear combinations of them, as $\gamma_1 e_1 + \gamma_2 e_2 + \cdots + \gamma_n e_n$, $\kappa_1 e_1 + \kappa_2 e_2 + \cdots + \kappa_n e_n$. Any set of n independent linear combinations of the units $e_1, e_2, \ldots, e_n$ may be regarded as constituting a set of fundamental units, since all numbers of the form $\alpha_1 e_1 + \alpha_2 e_2 + \cdots + \alpha_n e_n$ may be expressed linearly in terms of them.

6. GRAPHICAL REPRESENTATION OF NUMBERS. THE VARIABLE.

40. Correspondence between the Real Number-System and the Points of a Line. Let a right line be chosen, and on it a fixed point, to be called the null-point; also a fixed unit for the measurement of lengths.

Lengths may be measured on this line either from left to right or from right to left, and equal lengths measured in opposite directions, when added, annul each other; opposite algebraic signs may, therefore, be properly attached to them. Let the sign $+$ be attached to lengths measured to the right, the sign $-$ to lengths measured to the left.

The entire system of real numbers may be represented by the points of the line, by taking to correspond to each number that point whose distance from the null-point is represented by the number. For, as we proceed to demonstrate, the distance of every point of the line from the null-point, measured in terms of the fixed unit, is a real number; and we may assume that for each real number there is such a point.

1. *The distance of any point on the line from the null-point is a real number.*

Let any point on the line be taken, and suppose the segment of the line lying between this point and the null-point to contain the unit line α times, with a remainder d_1, this remainder to contain the tenth part of the unit line β times, with a remainder d_2, d_2 to contain the hundredth part of the unit line γ times, with a remainder d_3, etc.

The sequence of rational numbers thus constructed, viz., $\alpha, \alpha.\beta, \alpha.\beta\gamma, \ldots$ (adopting the decimal notation) is regular; for the difference between its μth term and each succeeding term is less than $\frac{1}{10^{\mu-1}}$, a fraction which may be made less than any assignable number by taking μ great enough; and, by construction, this number represents the distance of the point under consideration from the null-point.

By the convention made respecting the algebraic signs of lengths this number will be positive when the point lies to the right of the null-point, negative when it lies to the left.

2. *Corresponding to every real number there is a point on the line, whose distance and direction from the null-point are indicated by the number.*

(*a*) If the number is rational, we can construct the point.

For every rational number can be reduced to the form of a simple fraction. And if $\frac{\alpha}{\beta}$ denote the given number, when thus expressed, to find the corresponding point we have only to lay off the βth part of the unit segment α times along the line, from the null-point to the right, if $\frac{\alpha}{\beta}$ is positive, from the null-point to the left, if $\frac{\alpha}{\beta}$ is negative.

(*b*) If the number is irrational, we usually cannot construct the point, or even prove that it exists.

But let $\mathbf{a}$ denote the number, and $\alpha_1, \alpha_2, \ldots, \alpha_n, \ldots$ any regular sequence of rationals which defines it, so that α_n will approach $\mathbf{a}$ as limit when n is indefinitely increased.

Then, by (*a*), there is a sequence of points on the line corresponding to this sequence of rationals. Call this sequence of points $A_1, A_2, \cdots, A_n, \cdots$. It has the property that the length of the segment A_nA_{n+m} will approach 0 as limit when n is indefinitely increased.

When α_n is made to run through the sequence of values $\alpha_1, \alpha_2, \ldots$, the corresponding point A_n will run through the sequence of positions $A_1, A_2, \cdots$. And we *assume* that just as there is in the real system a definite number **a** which α_n is approaching as a limit, so also is there on the line a definite point **A** which A_n approaches as limit. It is this point **A** which we make correspond to **a**.

Of course there are infinitely many regular sequences of rationals $\alpha_1, \alpha_2, \ldots$ defining **a**, and as many sequences of corresponding points $A_1, A_2, \cdots$. We assume that the limit point **A** is the same for all these sequences.

41. The Continuous Variable. The relation of one-to-one correspondence between the system of real numbers and the points of a line is of great importance both to geometry and to algebra. It enables us, on the one hand, to express geometrical relations numerically, on the other, to picture complicated numerical relations geometrically. In particular, algebra is indebted to it for the very useful notion of the continuous variable.

One of our most familiar intuitions is that of continuous motion.

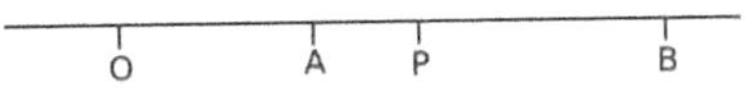

Suppose the point P to be moving continuously from A to B along the line OAB; and let **a**, **b**, and **x** denote the lengths of the segments OA, OB, and OP respectively, O being the null-point.

It will then follow from our assumption that the segment AB contains a point for every number between **a** and **b**, that as P moves continuously from A to B, **x** may be regarded as increasing from the value **a** to the value **b** through all intermediate values. To indicate this we call **x** a *continuous variable.*

42. Correspondence between the Complex Number-System and the Points of a Plane. The entire system of complex numbers may be represented by the points of a plane, as follows:

In the plane let two right lines $X'OX$ and $Y'OY$ be drawn intersecting at right angles at the point O.

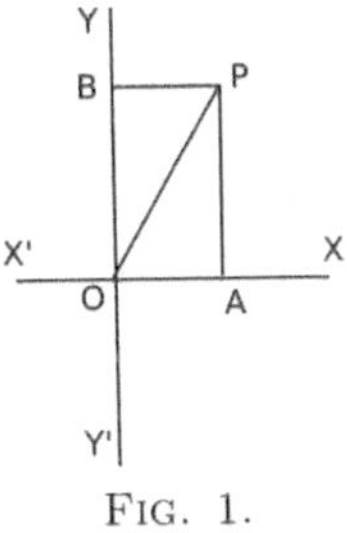

FIG. 1.

Make $X'OX$ the "axis" of real numbers, using its points to represent real numbers, after the manner described in § 40, and make $Y'OY$ the axis of pure imaginaries, representing ib by the point of OY whose distance from O is b when b is positive, and by the corresponding point of OY' when b is negative.

The point taken to represent the complex number $a + ib$ is P, constructed by drawing through A and B, the points which represent a and ib, parallels to $Y'OY$ and $X'OX$, respectively.

The correspondence between the complex numbers and the points of the plane is a one-to-one correspondence. To every point of the plane there is a complex number corresponding, and but one, while to each number there corresponds a single point of the plane.[1]

If the point P be made to move along any curve in its plane, the corresponding number x may be regarded as changing through a continuous system of complex values, and is called a *continuous complex variable.* (Compare § 41.)

43. Modulus. The length of the line OP (Fig. 1), *i. e.* $\sqrt{a^2+b^2}$, is called the *modulus* of $a+ib$. Let it be represented by ρ.

44. Argument. The angle XOP made by OP with the positive half of the axis of real numbers is called the *angle* of $a+ib$, or its *argument.* Let its numerical measure be represented by θ.

The angle is always to be measured "counter-clockwise" from the positive half of the axis of real numbers to the modulus line.

45. Sine. The ratio of PA, the perpendicular from P to the axis of real numbers, to OP, *i. e.* $\frac{b}{\rho}$, is called the *sine* of θ, written $\sin\theta$.

$\sin\theta$ is by this definition positive when P lies above the axis of real numbers, negative when P lies below this line.

46. Cosine. The ratio of PB, the perpendicular from P to the axis of imaginaries, to OP, *i. e.* $\frac{a}{\rho}$, is called the *cosine* of *theta*, written $\cos\theta$.

$\cos\theta$ is positive or negative according as P lies to the right or the left of the axis of imaginaries.

47. Theorem. *The expression of $a+ib$ in terms of its modulus and angle is* $\rho(\cos\theta+i\sin\theta)$.

For by § 46 $$\frac{a}{\rho}=\cos\theta, \therefore a=\rho\cos\theta;$$

and by § 45, $$\frac{b}{\rho}=\sin\theta, \therefore b=\rho\sin\theta.$$

Therefore $$a+ib=\rho(\cos\theta+i\sin\theta).$$

The factor $\cos\theta+i\sin\theta$ has the same sort of geometrical meaning as the algebraic signs + and −, which are indeed but particular cases of it: it indicates the *direction* of the point which represents the number from the null-point.

It is the other factor, the modulus ρ, the distance from the null-point of the point which corresponds to the number, which indicates the "absolute value" of the number, and may represent it when compared numerically with other numbers (§ 37),—that one of two numbers being numerically the greater whose corresponding point is the more distant from the null-point.

48. Problem I. *Given the points P and P', representing $a+ib$ and $a'+ib'$ respectively; required the point representing $a+a'+i(b+b')$.*

The point required is P'', the intersection of the parallel to OP through P' with the parallel to OP' through P.

---

[1]A reality has thus been found to correspond to the hitherto uninterpreted symbol $a+ib$. But this reality has no connection with the reality which gave rise to arithmetic, the number of things in a group of distinct things, and does not at all lessen the purely symbolic character of $a+ib$ when regarded from the standpoint of that reality, the standpoint which must be taken in a purely arithmetical study of the origin and nature of the number concept.

The connection between the numbers $a+ib$ and the points of a plane is purely artificial. The tangible geometrical pictures of the relations among complex numbers to which it leads are nevertheless a valuable aid in the study of these relations.

For completing the construction indicated by the figure, we have $OD' = PE = DD''$, and therefore $OD'' = OD + OD'$; and similarly $P''D'' = PD + P'D'$.

COR. I. To get the point corresponding to $a - a' + i(b - b')$, produce OP' to P''', making $OP''' = OP'$, and complete the parallelogram OP, OP'''.

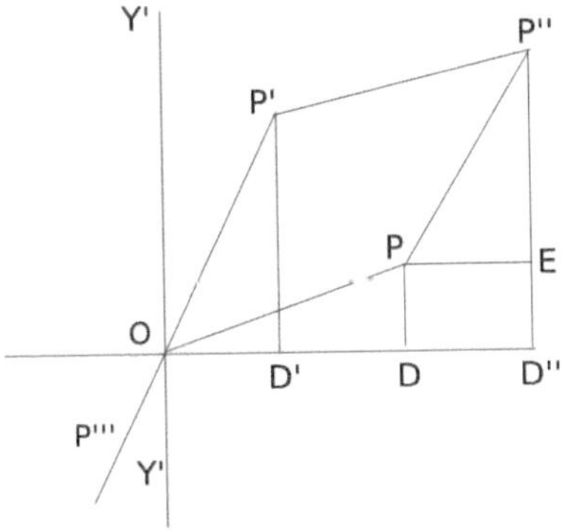

FIG. 2.

COR. II. *The modulus of the sum or difference of two complex numbers is less than (at greatest equal to) the sum of their moduli.*

For OP'' is less than $OP + PP''$ and, therefore, than $OP + OP$, unless O, P, P' are in the same straight line, when $OP'' = OP + OP'$. Similarly, PP', which is equal to the modulus of the difference of the numbers represented by P and P', is less than, at greatest equal to, $OP + OP'$.

49. Problem II. *Given P and P', representing $a + ib$ and $a' + ib'$ respectively; required the point representing $(a + ib)(a' + ib')$.*

$$\begin{aligned}
\text{Let}\quad a + ib &= \rho(\cos\theta + i\sin\theta), && \S\ 47\\
\text{and}\quad a' + ib' &= \rho'(\cos\theta' + i\sin\theta');\\
\text{then}\quad (a + ib)&(a' + ib')\\
&= \rho\rho'(\cos\theta + i\sin\theta)(\cos\theta' + i\sin\theta')\\
&= \rho\rho'[(\cos\theta\cos\theta' - \sin\theta\sin\theta')\\
&\qquad + i(\sin\theta\cos\theta' + \cos\theta\sin\theta')].\\
\text{But}\quad \cos\theta\cos\theta' &- \sin\theta\sin\theta' = \cos(\theta + \theta'),[1]\\
\text{and}\quad \sin\theta\cos\theta' &+ \cos\theta\sin\theta' = \sin(\theta + \theta').[1]
\end{aligned}$$

Therefore $(a + ib)(a' + ib') = \rho\rho'[\cos(\theta + \theta') + i\sin(\theta + \theta')]$; or, *The modulus of the product of two complex numbers is the product of their moduli, its argument the sum of their arguments.*

The required construction is, therefore, made by drawing through O a line making an angle $\theta + \theta'$ with OX, and laying off on this line the length $\rho\rho'$.

COR. I. Similarly the product of n numbers having moduli ρ, ρ', ρ'', $\cdots$ $\rho^{(n)}$

[1] For the demonstration of these, the so-called addition theorems of trigonometry, see Wells' Trigonometry, § 65, or any other text-book of trigonometry.

respectively, and arguments θ, θ', θ'', ... $theta^{(n)}$, is the number

$$\rho\rho'\rho''\cdots\rho^{(n)}[\cos(\theta+\theta'+\theta''+\cdots\theta^{(n)}) \\ +i\sin(\theta+\theta'+\theta''+\cdots\theta^{(n)})].$$

In particular, therefore, by supposing the n numbers equal, we may infer the theorem

$$[\rho(\cos\theta+i\sin\theta)]^n=\rho^n(\cos n\theta+i\sin n\theta),$$

which is known as Demoivre's Theorem.

COR. II. From the definition of division and the preceding demonstration it follows that

$$\frac{a+ib}{a'+ib'}=\frac{\rho}{\rho'}[\cos(\theta-\theta')+i\sin(\theta-\theta')];$$

the construction for the point representing $\dfrac{a+ib}{a'+ib'}$ is, therefore, obvious.

50. Circular Measure of Angle. Let a circle of unit radius be constructed with the vertex of any angle for centre. The length of the arc of this circle which is intercepted between the legs of the angle is called the *circular measure* of the angle.

51. Theorem. *Any complex number may be expressed in the form $\rho e^{i\theta}$; where ρ is its modulus and θ the circular measure of its angle.*

It has already been proven that a complex number may be written in the form $\rho(\cos\theta+i\sin\theta)$, where ρ and θ have the meanings just given them. The theorem will be demonstrated, therefore, when it shall have been shown that

$$e^{i\theta}=\cos\theta+i\sin\theta.$$

If n be any positive integer, we have, by § 36 and the binomial theorem,

$$\left(1+\frac{i\theta}{n}\right)^n=1+n\frac{i\theta}{n}+\frac{n(n-1)}{2!}\frac{(i\theta)^2}{n^2} \\ +\frac{n(n-1)(n-2)}{3!}\frac{(i\theta)^3}{n^3}+\cdots \\ =1+i\theta+\frac{1-\frac{1}{n}}{2!}(i\theta)^2 \\ +\frac{\left(1-\frac{1}{n}\right)\left(1-\frac{2}{n}\right)}{3!}(i\theta)^3+\cdots.$$

Let n be indefinitely increased; the limit of the right side of this equation will be the same as that of the left.

But the limit of the right side is

$$1+i\theta+\frac{(i\theta)^2}{2!}+\frac{(i\theta)^3}{3!}+\ldots;\ \text{i. e. } e^{i\theta}.$$[2]

Therefore $e^{i\theta}$ is the limit of $\left(1+\dfrac{i\theta}{n}\right)^n$ as n approaches ∞.

To construct the point representing $\left(1+\dfrac{i\theta}{n}\right)^n$:

On the axis of real numbers lay off $OA=1$.

[2]This use of the symbol $e^{i\theta}$ will be fully justified in § 73.

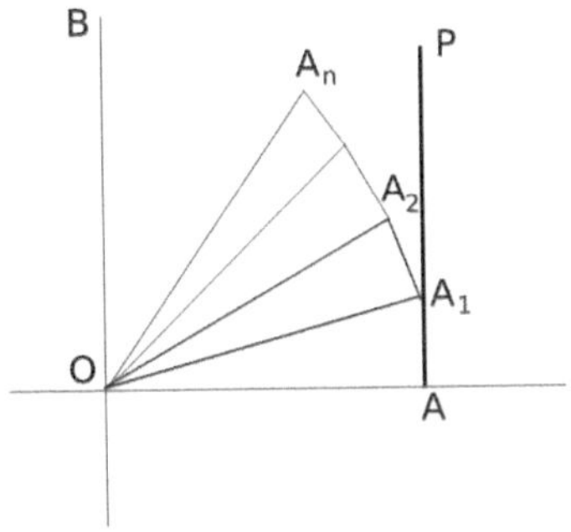

FIG. 3.

Draw AP equal to θ and parallel to OB, and divide it into n equal parts. Let AA_1 be one of these parts. Then A_1 is the point $1 + \dfrac{i\theta}{n}$.

Through A_1 draw A_1A_2 at right angles to OA_1 and construct the triangle OA_1A_2 similar to OAA_1.

A_2 is then the point $\left(1 + \dfrac{i\theta}{n}\right)^2$.

For $AOA_2 = 2AOA_1$;
and since $OA_2 : OA_1 :: OA_1 : OA$, and $OA = 1$,
the length $OA_2 =$ the square of length OA_1. (*see*§ 49)

In like manner construct A_3 to represent $\left(1 + \dfrac{i\theta}{n}\right)^3$, A_4 for $\left(1 + \dfrac{i\theta}{n}\right)^4$, $\cdots A_n$ for $\left(1 + \dfrac{i\theta}{n}\right)^n$.

Let n be indefinitely increased. The broken line $AA_1A_2 \cdots A_n$ will approach as limit an arc of length θ of the circle of radius OA and, therefore, its extremity, A_n, will approach as limit the point representing $\cos\theta + i\sin\theta$ (§ 47).

Therefore the limit of $\left(1 + \dfrac{i\theta}{n}\right)^n$ as n is indefinitely increased is $\cos\theta + i\sin\theta$.

But this same limit has already been proved to be $e^{i\theta}$.

Hence $\qquad e^{i\theta} = \cos\theta + i\sin\theta.$[3]

[3]Dr. F. Franklin, American Journal of Mathematics, Vol. VII, p. 376. Also Möbius, Collected Works, Vol. IV, p. 726.

7. THE FUNDAMENTAL THEOREM OF ALGEBRA.

52. The General Theorem. If

$$w = a_0z^n + a_1z^{n-1} + a_2z^{n-2} + \cdots + a_{n-1}z + a_n,$$

where n is a positive integer, and $a_0, a_1, \ldots, a_n$ any numbers, real or complex, independent of z, to each value of z corresponds a single value of w.

We proceed to demonstrate that conversely to each value of w corresponds a set of n values of z, *i. e.* that there are n numbers which, substituted for z in the polynomial $a_0z^n + a_1z^{n-1} + \cdots + a_n$, will give this polynomial any value, w_0, which may be assigned.

It will be sufficient to prove that there are n values of z which render $a_0z^n + a_1z^{n-1} + \cdots + a_n$ equal to 0, inasmuch as from this it would immediately follow that the polynomial takes any other value, w_0, for n values of z; viz., for the values which render the polynomial of the same degree, $a_0z^n + a_1z^{n-1} + \cdots + (a_n - w_0)$, equal to 0.

53. Root of an Equation. A value of z for which $a_0z^n + a_1z^{n-1} + \cdots + a_n$ is 0 is called a root of this polynomial, or more commonly a root of the *algebraic equation*

$$a_0z^n + a_1z^{n-1} + \cdots + a_n = 0.$$

54. Theorem. *Every algebraic equation has a root.*

Given $w = a_0z^n + a_1z^{n-1} + \cdots + a_n$.

Let $|w|$ denote the modulus of w. We shall assume, though this can be proved, that among the values of $|w|$ corresponding to all possible values of z there is a *least* value, and that this least value corresponds to a finite value of z.

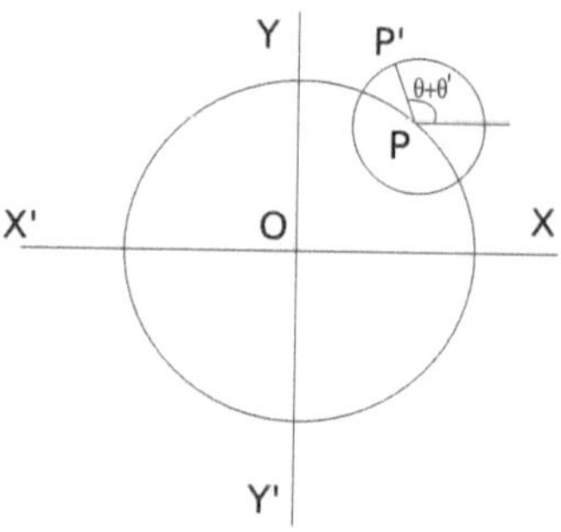

FIG. 4.

Let $|w_0|$ denote this least value of $|w|$, and z_0 the value of z to which it corresponds. Then $|w_0| = 0$.

For if not, w_0 will be represented in the plane of complex numbers by some point P distinct from the null-point O.

Through P draw a circle having its centre in the null-point O. Then, by the hypothesis made, no value can be given z which will bring the corresponding w-point within this circle.

But the w-point *can be brought within this circle.*

For, z_0 and w_0 being the values of z and w which correspond to P, change z by adding to z_0 a small increment δ, and let Δ represent the consequent change in w. Then Δ is defined by the equation

$$(w_0 + \Delta) = a_0(z_0 + \delta)^n + a_1(z_0 + \delta)^{n-1} \\ + a_2(z_0 + \delta)^{n-2} + \cdots + a_{n-1}(z_0 + \delta) + a_n.$$

On applying the binominal theorem and arranging the terms with reference to powers of δ, the right member of this equation becomes

$$a_0 z_0^n + a_1 z_0^{n-1} + \cdots + a_{n-1} z_0 + a_n \\ + [n a_0 z_0^{n-1} + (n-1) a_1 z_0^{n-2} + \cdots + a_{n-1}]\delta \\ + \text{ terms involving } \delta^2, \delta^3, \text{etc.}$$

But $$w_0 = a_0 z_0^n + a_1 z_0^{n-1} + \cdots + a_{n-1} z_0 + a_n.$$

$$\therefore \Delta = [n a_0 z_0^{n-1} + (n-1) a_1 z_0^{n-2} + \cdots + a_{n-1}]\delta \\ + \text{ terms involving } \delta^2, \delta^3, \text{etc.}$$

Let $\rho'(\cos\theta' + i\sin\theta')$ be the complex number

$$n a_0 z_0^{n-1} + (n-1) a_1 z_0^{n-2} + \cdots + a_{n-1},$$

expressed in terms of its modulus and angle, and

$$\rho(\cos\theta + i\sin\theta)$$

the corresponding expression for δ. Then

$$\Delta = \rho'(\cos\theta' + i\sin\theta') \times \rho(\cos\theta + i\sin\theta) \\ + \text{ terms involving } \rho^2, \rho^3, \text{etc.} \\ = \rho\rho'[\cos(\theta + \theta') + i\sin(\theta + \theta')] \\ + \text{ terms involving } \rho^2, \rho^3, \text{etc.} \qquad \S\ 49.$$

The point which represents $\rho\rho'[\cos(\theta + \theta') + i\sin(\theta + \theta')]$ for any particular value of ρ can be made to describe a circle of radius $\rho\rho'$ about the null-point by causing θ to increase continuously from 0 to 4 right angles.

In the same circumstances the point representing

$$w_0 + \rho\rho'[\cos(\theta + \theta') + i\sin(\theta + \theta')]$$

will describe an equal circle about the point P and, therefore, come within the circle OP.

But by taking ρ small enough, Δ may be made to differ as little as we please from $\rho\rho'[\cos(\theta+\theta') + i\sin(\theta+\theta')]$,[1] and, therefore, the curve traced out by P' (which represents $w_0 + \Delta$, as θ runs through its cycle of values), to differ as little as we please from the circle of centre P and radius $\rho\rho'$.

Therefore by assigning proper values to ρ and θ, the w-point (P') may be brought within the circle OP.

The w-point nearest the null-point must therefore be the null-point itself.[2]

55. Theorem. *If α be a root of $a_0z^n + a_1z^{n-1} + \cdots + a_n$, this polynomial is divisible by $z - a$.*

For divide $a_0z^n + a_1z^{n-1} + \cdots + a_n$ by $z - a$, continuing the division until z disappears from the remainder, and call this remainder R, the quotient Q, and, for convenience, the polynomial $f(z)$.

Then we have immediately

$$f(z) = (z - \alpha)Q + R,$$

holding for all values of z.

Let z take the value α; then $f(z)$ vanishes, as also the product $(z - \alpha)Q$.

Therefore when $z = \alpha$, $R = 0$, and being independent of z it is hence always 0.

56. The Fundamental Theorem. *The number of the roots of the polynomial $a_0z^n + a_1z^{n-1} + \cdots + a_n$ is n.*

For, by § 54, it has at least one root; call this α; then, by § 55, it is divisible by $z - \alpha$, the degree of the quotient being $n - 1$.

Therefore we have

$$a_0z^n + a_1z^{n-1} + \cdots + a_n = (z - \alpha)(a_0z^{n-1} + b_1z^{n-2} + \cdots + b_{n-1}).$$

Again, by § 54, the polynomial $a_0z^{n-1} + b_1z^{n-2} + \cdots + b_{n-1}$ has a root; call this β, and dividing as before, we have

$$a_0z^n + a_1z^{n-1} + \cdots + a_n = (z - \alpha)(z - \beta)(\alpha_1z^{n-2} + c_1z^{n-3} + \cdots + c_{n-2}).$$

Since the degree of the quotient is lowered by 1 by each repetition of this process, $n - 1$ repetitions reduce it to the first degree, or we have

$$a_0z^n + a_1z^{n-1} + \cdots + a_n = a_0(z - \alpha)(z - \beta)(z - \gamma)\cdots(z - \nu),$$

a product of n factors, each of the first degree.

Now a product vanishes when one of its factors vanishes (§ 36, 3, Cor.), and the factor $z - \alpha$ vanishes when $z = \alpha$, $z - \beta$ when $z = \beta, \ldots, z - \nu$ when $z = \nu$. Therefore $a_0z^n + a_0z^{n-1} + \cdots + a_n$ vanishes for the n values, $\alpha, \beta, \gamma, \cdots \nu$, of z.

Furthermore, a product cannot vanish unless one of its factors vanishes (§ 36, 4, Cor.), and not one of the factors $z - \alpha, z - \beta, \ldots, z - \nu$, vanishes unless z equals one of the numbers $\alpha, \beta, \cdots \nu$.

The polynomial has therefore n and but n roots.

The theorem that the number of roots of an algebraic equation is the same as its degree is called the fundamental theorem of algebra.

---

[1] In the series $A\rho + B\rho^2 + C\rho^3 +$ etc., the ratio of all the terms following the first to the first, *i. e.*

$$\frac{B\rho^2 + c\rho^3 + \text{ etc.}}{A\rho}, = \rho \times \frac{B + C\rho + \text{ etc.}}{A};$$

which by taking ρ small enough may evidently be made as small as we please.

[2] In the above demonstration it is assumed that the coefficient of δ is not 0. If it be 0, let $A\delta^r$ denote the first term of Δ which is not 0. If $A = \rho''(\cos\theta'' + i\sin\theta'')$, we then have

$$\Delta = \rho''\rho^r[\cos(r\theta + \theta'') + i\sin(r\theta + \theta'')] + \text{ terms in } \theta^{r+1}, \ldots b,$$

from which the same conclusion follows as above.

8. INFINITE SERIES.

57. Definition. Any operation which is the limit of additions indefinitely repeated produces an infinite series. We are to determine the conditions which an infinite series must fulfil to represent a number.

If the terms of a series are real numbers, it is called a *real series*; if complex, a *complex series*.

8.1 REAL SERIES.

58. Sum. Convergence. Divergence. An infinite series

$$a_1 + a_2 + a_3 + \cdots + a_n + \cdots$$

represents a number or not, according as the sequence

$$s_1, s_2, s_3, \ldots s_m, s_{m+1}, \ldots s_{m+n}, \cdots,$$

$$\text{where} \qquad s_1 = a_1, s_2 = a_1 + a_2, \cdots, s_i = a_1 + a_2 + \cdots a_i,$$

is *regular* or not.

If $s_1, s_2, \cdots$, be a regular sequence, the number which it defines, or $\lim_{n \doteq \infty}(s_n)$, is called the *sum* of the infinite series

$$a_1 + a_2 + a_3 + \cdots + a_n + \cdots,$$

and the series is said to be *convergent*.

If s_1, s_2, be not a regular sequence, s_n either transcends any finite value whatsoever, as n is indefinitely increased, or while remaining finite becomes altogether indeterminate. The infinite series then has no sum, and is said to be *divergent*.

The series $1 + 1 + 1 + \cdots$ and $1 - 1 + 1 - 1 + \cdots$ are examples of these two classes of divergent series.

A divergent series cannot represent a number.

59. General Test of Convergence. From these definitions and § 27 it immediately follows that:

The infinite series $a_1 + a_2 + \cdots + a_m + \cdots$ is convergent when m may be so taken that the differences $s_{m+n} - s_m$ are numerically less than any assignable number δ for all values of n, where s_m and s_{m+n} are the sum of the first m and of the first $m + n$ terms of the series respectively.

If these conditions be not fulfilled, the series is divergent.

The limit of the nth term of a convergent series is 0; for the condition of convergence requires that by taking m great enough, $s_{m+1} - s_m$, *i. e.* a_{m+1}, may be found less than any assignable number. But it is not to be assumed conversely that a series is convergent, if the limit of its nth term is 0; other conditions have also to be fulfilled, $s_{m+n} - s_m$ must be less than δ for *all* values of n.

Thus the limit of the nth term of the series $1 + \frac{1}{2} + \frac{1}{3} + \cdots$ is 0; but, as will presently be shown, this is a divergent series.

60. Absolute Convergence. It is important to distinguish between convergent series which remain convergent when all the terms are given the same algebraic signs and convergent series which become divergent on this change of signs. Series of the first class are said to be *absolutely* convergent; those of the second class, only *conditionally* convergent.

Absolutely convergent series have the character of ordinary sums; i. e. the order of the terms may be changed without altering the sum of the series.

For consider the series $a_1 + a_2 + a_3 + \cdots$ supposed to be absolutely convergent and to have the sum S, when the terms are in the normal order of the indices.

It is immediately obvious that no change can be made in the sum of the series by interchanging terms with finite indices; for n may be taken greater than the index of any of the interchanged terms. Then S_n has not been affected by the change, since it is a finite sum and it is immaterial in what order the terms of a finite sum are added; and as for the rest of the series, no change has been made in the order of its terms.

But $a_1 + a_2 + a_3 + \cdots$ may be separated into a number of infinite series, as, for instance, into the series $a_1 + a_3 + a_5 + \cdots$ and $a_2 + a_4 + a_6 + \cdots$, and these series summed separately. Let it be separated into l such series, the sums of which—they must all be absolutely convergent, as being parts of an absolutely convergent series—are $S^{(1)}, S^{(2)}, \cdots S^{(l)}$, respectively; it is to be proven that

$$S = S^{(1)} + S^{(2)} + S^{(3)} + \cdots + S^{(l)}.$$

Let $S_m^{(1)}, S_m^{(2)}, \cdots$ be the sums of the first m terms of the series $S^{(1)}, S^{(2)}, \cdots$, respectively.

Then, by the hypothesis that the series $a_1 + a_2 + \cdots$ is absolutely convergent, m may be taken so large that the sum

$$S_{m+n}{}^{(1)} + S_{m+n}{}^{(2)} + \cdots + S_{m+n}{}^{(l)}$$

shall differ from S by less than any assignable number δ for all values of n; therefore the limit of this sum is S.

But again, n may be so taken that $S_{m+n}{}^{(1)}$ shall differ from $S^{(1)}$ by less than $\frac{\delta}{l}$, $S_{m+n}{}^{(2)}$ from $S^{(2)}$ by less than $\frac{\delta}{l}, \ldots$; and therefore the sum $S_{m+n}{}^{(1)} + S_{m+n}{}^{(2)} + \cdots + S_{m+n}{}^{(l)}$ from $S^{(1)} + S^{(2)} + \cdots + S^{(l)}$ by less than $\left(\frac{\delta}{l}\right) l$; *i. e.* by less than δ. Hence the limit of this sum is $S^{(1)} + S^{(2)} + \cdots + S^{(l)}$.

Therefore S and $S^{(1)} + S^{(2)} + \cdots + S^{(l)}$ are limits of the same finite sum and hence equal. (We omit the proof for the case l infinite.)

61. Conditional Convergence. On the other hand, *the terms of a conditionally convergent series can be so arranged that the sum of the series may take any real value whatsoever.*

In a conditionally convergent series the positive and the negative terms each constitute a divergent series having 0 for the limit of its last term.

If, therefore, C be any positive number, and S_n be constructed by first adding positive terms (beginning with the first) until their sum is greater than C, to these negative terms until their sum is again less than C, then positive terms till the sum is again greater than C, and so on indefinitely; the limit of S_n, as n is indefinitely increased, is C.

62. Special Tests of Convergence. 1. *If each of the terms of a series $a_1 + a_2 + \cdots$ be numerically less than (at greatest equal to) the corresponding term of an absolutely convergent series, or if the ratio of each term of $a_1 + a_2 + \cdots$ to the* *corresponding term of an absolutely convergent series never exceed some finite number C, the series $a_1 + a_2 + \cdots$ is absolutely convergent.*

If, on the other hand, each term of $a_1 + a_2 + \cdots$ be numerically greater than (at the lowest equal to) the corresponding term of a divergent series, or if the ratio of each term of $a_1 + a_2 + \cdots$ to the corresponding term of a divergent series be never numerically less than some finite number C', different from 0, the series $a_1 + a_2 + \cdots$ is divergent.

2. *The series $a_1 - a_2 + a_3 - a_4 + \cdots$, the terms of which are alternately positive and negative, is convergent, if after some term a_i each term be numerically less or, at least, not greater than the term which immediately precedes it, and the limit of a_n, as n is indefinitely increased, be 0.*

For here

$$s_{m+n} - s_m = (-1)^m [a_{m+1} - a_{m+2} + \cdots (-1)^{n-1} a_{m+n}]$$

The expression within brackets may be written in either of the forms

$$(a_{m+1} - a_{m+2}) + (a_{m+3} - a_{m+4}) + \cdots \tag{1}$$

or

$$a_{m+1} - (a_{m+2} - a_{m+3}) - \cdots \tag{2}$$

It is therefore positive, (1), and less than a_{m+1}, (2); and hence by taking m large enough, may be made numerically less than any assignable number whatsoever.

The series $1 - \frac{1}{2} + \frac{1}{3} - \frac{1}{4} + \cdots$ is, by this theorem, convergent.

3. *The series $1 + \frac{1}{2} + \frac{1}{3} + \frac{1}{4} + \cdots$ is divergent.*

For the first 2^λ terms after the first may be written

$$\frac{1}{2} + \left(\frac{1}{2+1} + \frac{1}{2+2}\right) + \left(\frac{1}{2^2+1} + \frac{1}{2^2+2} + \frac{1}{2^2+3} + \frac{1}{2^2+2^2}\right) + \cdots$$
$$+ \left(\frac{1}{2^{\lambda-1}+1} + \frac{1}{2^{\lambda-1}+2} + \cdots \frac{1}{2^{\lambda-1}+2^{\lambda-1}}\right),$$

where, obviously, each of the expressions within parentheses is greater than $\frac{1}{2}$.

The sum of the first 2^λ terms after the first is therefore greater than $\frac{\lambda}{2}$, and may be made to exceed any finite quantity whatsoever by taking λ great enough.

This series is commonly called the harmonic series.

By a similar method of proof it may be shown that the series $1 + \frac{1}{2^p} + \frac{1}{3^p} + \cdots$ is convergent if $p > 1$.

Here, $$\frac{1}{2^p} + \frac{1}{3^p} < \frac{2}{2^p}, \; \frac{1}{4^p} + \frac{1}{5^p} + \frac{1}{6^p} + \frac{1}{7^p} < \frac{4}{4^p}, \; i.\ e. < \left(\frac{2}{2^p}\right)^2 \cdots,$$

and the sum of the series is, therefore, less than that of the decreasing geometric series $1 + \frac{2}{2^p} + \left(\frac{2}{2^p}\right)^2 + \cdots$.

The series $1 + \frac{1}{2^p} + \frac{1}{3^p} + \cdots$ is divergent if $p < 1$, the terms being then greater than the corresponding terms of

$$1 + \frac{1}{2} + \frac{1}{3} + \cdots.$$

4. *The series $a_1 + a_2 + a_3 + \cdots$ is absolutely convergent if after some term of finite index, a_i, the ratio of each term to that which immediately precedes it be numerically less than 1 and, as the index of the term is indefinitely increased, approach a limit which is less than 1; but divergent, if this ratio and its limit be greater than 1.*

For—to consider the first hypothesis—suppose that after the term a_i this ratio is always less than α, where α denotes a certain positive number less than 1.

$$\text{Then,}\quad \frac{a_{i+1}}{a_i} \leqq \alpha, \ \therefore\ a_{i+1} \leqq a_i\alpha;$$

$$\frac{a_{i+2}}{a_{i+1}} \leqq \alpha, \ \therefore\ a_{i+2} \leqq a_{i+1}\alpha \leqq a_i\alpha^2.$$

$$\cdot\quad\cdot\quad\cdot\quad\cdot\quad\cdot\quad\cdot\quad\cdot$$

$$\frac{a_{i+k}}{a_{i+(k-1)}} \leqq \alpha, \ \therefore\ a_{i+k} \leqq a_{i+(k-1)}\alpha \leqq \cdots \leqq a_i\alpha^k.$$

$$\cdot\quad\cdot\quad\cdot\quad\cdot\quad\cdot\quad\cdot\quad\cdot$$

The given series is therefore $\leqq$

$$s_i + a_i[\alpha + \alpha^2 + \alpha^3 + \cdots \alpha^k + \cdots].$$

And this is an absolutely convergent series.

$$\begin{aligned}\text{For}\quad \alpha + \alpha^2 + \cdots \alpha^k + \cdots &= \lim_{n \doteq \infty} (\alpha + \alpha^2 + \cdots + \alpha^n)\\ &= \lim_{n \doteq \infty} \left(\frac{\alpha - \alpha^{n+1}}{1 - \alpha}\right)\\ &= \frac{\alpha}{1 - \alpha},\ \text{since } \alpha \text{ is a fraction.}\end{aligned}$$

The given series is therefore absolutely convergent, § 62, 1.

The same course of reasoning would prove that the series is divergent when after some term α_i the ratio of each term to that which precedes it is never less than some quantity, α, which is itself greater than 1.

When the limit of the ratio of each term of the series to the term immediately preceding it is 1, the series is sometimes convergent, sometimes divergent. The series considered in § 62, 3 are illustrations of this statement.

63. Limits of Convergence. An important application of the theorem just demonstrated is in determining what are called the limits of convergence of infinite series of the form

$$a_0 + a_1x + a_2x^2 + a_3x^3 + \cdots,$$

where x is supposed variable, but the coefficients a_0, a_1, etc., constants as in the preceding discussion. Such a series will be convergent for very small values of x, if the coefficients be all finite, as will be supposed, and generally divergent for very great values of x; and by the limits of convergence of the series are meant the values of x for which it ceases to be convergent and becomes divergent.

By the preceding theorem the series will be *convergent* if the limit of the ratio of any term to that which precedes it be numerically less than 1; *i. e.* if

$$\lim_{n \doteq \infty} \left(\frac{a_{n+1} x^{n+1}}{a_n x^n} \right), \text{ or } \lim_{n \doteq \infty} \left(\frac{a_{n+1}}{a_n} x \right), < 1;$$

that is, *if x be numerically* $< \lim_{n \doteq \infty} \left(\frac{a_n}{a_{n+1}} \right)$; and *divergent, if x be numerically* $>$ $\lim_{n \doteq \infty} \left(\frac{a_n}{a_{n+1}} \right)$.

1. Thus the infinite series

$$a^m + ma^{m-1}x + \frac{m(m-1)}{2!} a^{m-2} x^2 + \cdots,$$

which is the expansion, by the binomial theorem, of $(a+x)^m$ for other than positive integral values of m, is convergent for values of x numerically less than a, divergent for values of x numerically greater than a.

For in this case

$$\lim_{n \doteq \infty} \left(\frac{a_n}{a_{n+1}} \right) = \lim_{n \doteq \infty} \left[a \times \frac{\frac{m(m-1)\cdots(m-n+1)}{(n)!}}{\frac{m(m-1)\cdots(m-n)}{(n+1)!}} \right]$$

$$= \lim_{n \doteq \infty} \left(a \times \frac{n+1}{m-n} \right)$$

$$= \lim_{n \doteq \infty} \left(\frac{a\left(1+\frac{1}{n}\right)}{-1+\frac{m}{n}} \right) = -a.$$

2. Again, the expansion of e^x, i. e. $1 + x + \frac{x^2}{2!} + \cdots$, is convergent for all finite values of x.

For here $\lim_{n \doteq \infty} \left(\frac{a_n}{a_{n+1}} \right) = \lim_{n \doteq \infty} \left(\frac{\frac{1}{(n)!}}{\frac{1}{(n+1)!}} \right) = \lim_{n \doteq \infty} (n+1) = \infty.$

The same is true for the series which is the expansion of a^x.

64. Operations on Infinite Series. 1. *The sum of two convergent series, $a_1 + a_2 + \cdots$ and $b_1 + b_2 + \cdots$, is the series $(a_1 + b_1) + (a_2 + b_2) + \cdots$; and their difference is the series $(a_1 - b_1) + (a_2 - b_2) + \cdots$.*

The sum of the series $a_1 + a_2 + \cdots$ is the number defined by $s_1, s_2, \cdots$, and the sum of the series $b_1 + b_2 + \cdots$ is the number defined by $t_1, t_2, \cdots$, where $s_i = a_1 + a_2 + \cdots + a_i$ and $t_i = b_1 + b_2 + \cdots + b_i$. The sum of the two series is therefore the number defined by $s_1 + t_1, s_2 + t_2, \cdots$, § 29, (1).

But if $S_i = (a_1 + b_1) + (a_2 + b_2) + \cdots + (a_i + b_i)$, we have $S_i = s_i + t_i$ for all values of i. This is immediately obvious for finite values of i, and there can be no difference between S_i and $s_i + t_i$ as i approaches ∞, since it would be a difference having 0 for its limit.

Therefore the number defined by $s_1 + t_1, s_2 + t_2, \cdots$, is the sum of the series $(a_1 + b_1) + (a_2 + b_2) + \cdots$.

2. *The product of two absolutely convergent series*

$$a_1 + a_2 + \cdots \text{ and } b_1 + b_2 + \cdots$$

is the series $a_1b_1 + (a_1b_2 + a_2b_1) + (a_1b_3 + a_2b_2 + a_3b_1) + \cdots$
$$+ (a_1b_n + a_2b_{n-1} + \cdots + a_{n-1}b_2 + a_nb_1) + \cdots.$$

Each set of terms within parentheses is to be regarded as constituting a single term of the product; and it will be noticed that the first of them consists of the one partial product in which the sum of the indices is 2, the second of all in which the sum of the indices is 3, etc.

By § 29, (3), the product of $a_1 + a_2 + \cdots$ by $b_1 + b_2 + \cdots$ is $\lim_{n \doteq \infty} (s_nt_n)$, where s_n and t_n represent the sums of the first n terms of $a_1+a_2+\cdots$, $b_1+b_2+\cdots$, respectively.

Suppose first that the terms of $a_1+a_2+\cdots$ and $b_1+b_2+\cdots$ are all positive. Then if S_n be the sum of the first n terms of $a_1b_1 + (a_1b_2 + a_2b_1) + \cdots$, and m represent $\frac{n}{2}$ when n is even and $\frac{n-1}{2}$ when n is odd,

$$\text{evidently} \quad s_nt_n > S_n > s_mt_m.$$

$$\text{But} \quad \lim_{n \doteq \infty} (s_nt_n) = \lim_{n \doteq \infty} (s_mt_m).$$

$$\text{Therefore} \quad \lim_{n \doteq \infty} (S_n) = \lim_{n \doteq \infty} (s_nt_n).$$

If the terms of $a_1 + a_2 + \cdots$, $b_1 + b_2 + \cdots$ be not all of the same sign, call the sums of the first n terms of the series got by making all the signs plus, s'_n and t'_n respectively; also S'_n, the sum of the first n terms of the series which is their product.

Then by the demonstration just given

$$\lim_{n \doteq \infty} (S'_n) = \lim_{n \doteq \infty} (s'_nt'_n);$$

but S_n always differs from s_nt_n by less than (at greatest by as much as) S'_n from $s'_nt'_n$; therefore, as before,

$$\lim_{n \doteq \infty} (S_n) = \lim_{n \doteq \infty} (s_nt_n).$$

3. The *quotient* of the series $a_0 + a_1x + \cdots$ by the series $b_0 + b_1x + \cdots$ (b_0 not 0) is a series of a similar form, as $c_0 + c_1x + \cdots$, which converges when $a_0 + a_1x + \cdots$ is absolutely convergent and $b_1x + \cdots$ is numerically less than b_0.

8.2 COMPLEX SERIES.

The terms *sum, convergent, divergent,* have the same meanings in connection with complex as in connection with real series.

65. General Test of Convergence. *A complex series, $a_1+a_2+\cdots$, is convergent when the modulus of $s_{m+n} - s_m$ may be made less than any assignable number δ by taking m great enough, and that for all values of n; divergent, when this condition is not satisfied.* See § 48, Cor. II; § 59.

66. Of Absolute Convergence. Let

$a_1 + a_2 + \cdots$ be a complex series,

and $A_1 + A_2 + \cdots$, the series of the moduli of its terms

If the series $A_1+A_2+\cdots$, be convergent, the series $a_1+a_2+\cdots$ will be convergent also.

For the modulus of the sum of a set of complex numbers is less than (at greatest equal to) the sum of their moduli (§ 48, Cor. II). By hypothesis, $S_{m+n} - S_m$ is less than any assignable number δ, when $S_m = A_1 + A_2 + \cdots + A_m$, etc.; much more must the modulus of $s_{m+n} - s_m$ be less than δ.

The converse of this theorem is not necessarily true; and a convergent series, $a_1 + a_2 + \cdots$, is said to be *absolutely* or only *conditionally* convergent, according as the series $A_1 + A_2 + \cdots$ is convergent or divergent.

67. The Region of Convergence of a Complex Series. *If the complex series $a_0 + a_1 z + a_2 z^2 + \cdots$ is convergent when $z = Z$, it is absolutely convergent for every value of z which is numerically less than Z, that is, it converges absolutely at every point within that circle in the plane of complex numbers which has the null-point for centre and passes through the point Z.*

For since the series $a_0 + a_1 Z + a_2 Z^2 + \cdots$ is convergent, its term $a_n Z^n$ approaches 0 as limit when n is indefinitely increased. It is therefore possible to find a real number M which is numerically greater than every term of this series.

Assign to z any value which is numerically less than Z, whose corresponding point, therefore, lies within the circle through the point Z.

For this value of z the terms of the series $a_0 + a_1 z + a_2 z^2 + \cdots$ will be numerically less than the corresponding terms of the series

$$M + M\frac{z}{Z} + M\left(\frac{z}{Z}\right)^2 + \cdots. \qquad (1)$$

For, since $a_n Z^n < M$, we have $a_n z^n < M\left(\frac{z}{Z}\right)^n$ numerically.

But the series (1) is absolutely convergent (§ 62, 4).

Therefore the given series $a_0+a_1z+a_2z^2+\cdots$ also is absolutely convergent for the value of z under consideration, that is, for all values of z whose corresponding points lie within the circle through the point Z.

NOTE. For other points than Z on the *circumference* of this circle through Z the series is not necessarily convergent.

Thus the series $1 + \frac{z}{2} + \frac{z^2}{3} + \cdots$ converges when $z = Z = -1$. But on the circle through the point -1, the point 1 also lies; and the series diverges when $z = 1$.

68. Theorem. The following is a theorem on which many of the properties of functions defined by series depend.

If the series $\quad a0 + a_1 z + a_2 z^2 + \cdots + a_n z^n + \cdots$

have a circle of convergence greater than the null-point itself, and z run through a regular sequence of values z_1, z_2, ... defining 0, the sum of all terms following the first, viz.,

$$a_1 z + a_2 z^2 + \cdots + a_n z^n + \cdots$$

will run through a sequence of values likewise regular and defining 0; or, the entire series may be made to differ as little as one chooses from its first term a_0.

The numbers z_1, z_2, ... are, of course, all supposed to lie within the circle of convergence, and for convenience, to be real. It will be convenient also to suppose $z_1 > z_2 > z_3$, etc.; i. e. that each is greater than the one following it.

Since $\quad a_0 + a_1 z + a_2 z^2 + \cdots + a_n z^n + \cdots$

converges absolutely for $z = z_1$, so also does

$$a_1 z + a_2 z^2 + \cdots + a_n z^n + \cdots,$$

and, therefore,

$$a_1 + a_2 z + \cdots + a_n z^{n-1} + \cdots.$$

Hence

$$A_1 + A_2 z_1 + \cdots + A_n z_1^{n-1} + \cdots$$

(where $A_i =$ modulus a_i) is convergent, and a number M can be found greater than its sum.

And since for $z = z_2, z_3, \ldots$ the individual terms of

$$A_1 + A_2 z + \cdots + A_n z^{n-1} + \cdots$$

are less than the corresponding terms of $A_1 + A_2 z_1 + \cdots + A_n z_1^{n-1} + \cdots$, this series and, therefore, *modulus*$(a_1 + a_2 z + \cdots)$ remain always less than M as z runs through the sequence of values $z_2, z_3, \cdots$.

Hence the values of *modulus*$(a_1 z + a_2 z^2 + \cdots)$ which correspond to $z = z_1, z_2 \ldots$ constitute a regular sequence defining 0, each term being numerically less than the corresponding term of the regular sequence $z_1 M, z_2 M, \ldots$ which defines 0.

COR. The same argument proves that if

$$a_m z^m + a_{m+1} z^{m+1} + \cdots,$$

or

$$z^m (a_m + a_{m+1} z + \cdots),$$

be the sum of all terms of the series from the $(m+1)$th on, the series $a_m + a_{m+1} z + \cdots$ can be made to differ as little as one may please from its first term a_m.

69. Operations on Complex Series. The definitions of *sum*, *difference*, and *product* of two convergent complex series are the same as those already given for real series, viz.:

1. *The sum of two convergent series,* $a_1 + a_2 + \cdots$ *and* $b_1 + b_2 + \cdots$, *is the series* $(a_1 + b_1) + (a_2 + b_2) + \cdots$; *their difference, the series* $(a_1 - b_1) + (a_2 - b_2) + \cdots$.

For if

$$s_i = a_1 + a_2 + \cdots + ai \text{ and } t_i = b_1 + b_2 + \cdots + bi,$$

$$\text{modulus } [(s_{m+n} \pm t_{m+n}) - (s_m \pm t_m)]$$

$$60mu \leq \text{ modulus } (s_{m+n} - s_m) + \text{ modulus } (t_{m+n} - t_m),$$

and may, therefore, be made less than any assignable number by taking m great enough. The theorem therefore follows by the reasoning of § 64, 1.

2. *The product of two absolutely convergent series,*

$$a_1 + a_2 + a_2 + \cdots \quad \textit{and } b_1 + b_2 + b_3 + \cdots,$$

is the series $a_1 b_1 + (a_1 b_2 + a_2 b_1) + (a_1 b_3 + a_2 b_2 + a_3 b_1) \cdots$.

For, letting $S_i = A_1 + A_2 + \cdots + A_i$ and $T_i = B_1 + B_2 + \cdots + B_i$, where A_i, B_i, are the moduli of a_i, b_i, respectively, and representing by σ_n the sum of the first n terms of the series

$$a_1 b_1 + (a_1 b_2 + a_2 b_1) + \cdots$$

and by Σ_n sum of the first n terms of the series

$$A_1B_1 + (A_1B_2 + A_2B_1) + \cdots,$$

we have

$$\text{modulus } (s_n t_n - \sigma_n) \leq S_n T_n - \Sigma_n.$$

But the limit of the right member of this inequality (or equation) is 0 (§ 64, 2); therefore

$$\lim_{n \doteq \infty} (\sigma_n) = \lim_{n \doteq \infty} (s_n t_n).$$

9. THE EXPONENTIAL AND LOGARITHMIC FUNCTIONS. UNDETERMINED COEFFICIENTS. INVOLUTION AND EVOLUTION. THE BINOMIAL THEORAM

70. Function. A variable w is said to be a *function* of a second variable z for the area A of the z-plane (§42), when to the z belonging to every point of A there corresponds a determinate value or set of values of w.

Thus if $w = 2z$, w is a function of z. For when $z = 1$, $w = 2$; when $z = 2$, $w = 4$; and there is in like manner a determinate value of w for every value of z. In this case A is coextensive with the entire z-plane.

Similarly w is a function of z, if

$$w = a_0 + a_1 z + a_2 z^2 + \ldots + a_n z^n + \ldots,$$

so long as this infinite series is convergent, *i. e.* for the portion of the z-plane bounded by a circle having the null-point for centre, and for radius the modulus of the smallest value of z for which the series diverges.

It is customary to use for w when a function of z the symbol $f(z)$, read "function z."

71. Functional Equation of the Exponential Function. For positive integral values of z and t, $a^z \cdot a^t = a^{z+t}$. The question naturally suggests itself, is there a function of z which will satisfy the condition expressed by this equation, or the "functional equation" $f(z)f(t) = f(z+t)$, for *all* values of z and t?

We proceed to the investigation of this question and another which it suggests, not only because they lead to definitions of the important functions a^z and $\log_a z$ for complex values of a and z, and so give the operations of involution, evolution, and the taking of logarithms the perfectly general character already secured to the four fundamental operations,—but because they afford simple examples of a large class of mathematical investigations.[1]

72. Undetermined Coefficients. In investigations of this sort, the method commonly used in one form or another is that of *undetermined coefficients.* This method consists in assuming for the function sought an expression involving a series of unknown but constant quantities—coefficients,—in substituting this expression in the equation or equations which embody the conditions which the function must satisfy, and in so determining these unknown constants that these equations shall be *identically* satisfied, that is to say, satisfied for all values of the variable or variables.

The method is based on the following theorem, called "the theorem of undetermined coefficients," viz.:

---

[1] An application of the principle of permanence (§12) is involved in the use of functional equations to define functions. The equation $a^z a^t = a^{z+t}$, for instance, only becomes a functional equation when its *permanence is assumed* for other values of z and t than those for which it has been actually demonstrated.

In this respect the methods of definition of the negative and the fraction on the one hand, and the functions a^z, $\log_a z$, on the other, are identical; but, while the equation $(a-b)+b = a$ itself served as definition of $a - b$, there being no simpler symbols in terms of which $a - b$ could be expressed, from the equation $a^z a^t = a^{z+t}$ a series (§ 73, (4)) may be deduced which defines a^z in terms of numbers of the system $a + ib$.

If the series $A + Bz + Cz^2 + \cdots$ be equal to the series $A' + B'z + C'z^2 + \cdots$ for all values of z which make both convergent, and the coefficients be independent of z, the coefficients of like powers of z in the two are equal.

For, since

$$A + Bz + Cz^2 + \cdots = A' + B'z + C'z^2 + \cdots,$$
$$A - A' + (B - B')z + (C - C')z^2 + \cdots = 0$$

throughout the circle of convergence common to the two given series (§§ 67, 69, 1).

And being convergent within this circle, the series

$$A - A' + (B - B')z + (C - C')z^2 + \cdots$$

can be made to differ as little as we please from its first term, $A - A'$ (§ 68).

$$\therefore A - A' = 0 \text{ (§ 30, Cor.), or } A = A'.$$

Therefore

$$(B - B')z + (C - C')z^2 + \cdots = 0$$

throughout the common circle of convergence, and hence (at least, for values of z different from 0)

$$B - B' + (C - C')z + \cdots = 0$$

Therefore by the reasoning which proved that

$$A - A' = 0, \; B - B' = 0, \text{ or } B = B'.$$

In like manner it may be proved that $C = C'$, $D = D'$, etc.

COR. *If* $\quad A + Bz + Ct + Dz^2 + Ezt + Ft^2 + \cdots$

$$= A' + B'z + C't + D'z^2 + E'zt + F't^2 + \cdots$$

for all values of z and t which make both series convergent, and z be independent of t, and the coefficients independent of both z and t, the coefficients of like powers of z and t in the two series are equal.

For, arrange both series with reference to the powers of either variable. The coefficients of like powers of this variable are then equal, by the preceding theorem. These coefficients are series in the other variable, and by applying the theorem to each equation between them the corollary is demonstrated.

73. The Exponential Function. To apply this method to the case in hand, assume

$$f(z) = A_0 + A_1 z + A_2 z^2 + \cdots + A_n z^n + \cdots,$$

and determine whether values of the coefficients A_i can be found capable of satisfying the "functional equation,"

$$f(z)f(t) = f(z + t), \tag{1}$$

for all values of z and t.

On substituting in this equation, we have, for all values of z and t for which the series converge,

$$(A_0 + A_1 z + A_2 z^2 + \cdots A_n z^n + \cdots)(A_0 + A_1 t + A_2 t^2 + \cdots A_n t^n + \cdots)$$
$$= A_0 + A_1(z + t) + A_2(z + t)^2 + \cdots A_n(z + t)^n + \cdots;$$

or, expanding and arranging the terms with reference to the powers of z and t,

$$A_0A_0 + A_1A_0z + A_0A_1t + A_2A_0z^2 + A_1A_1zt + A_0A_2t^2 + \cdots$$
$$+ A_nA_0z^n + A_{n-1}A_1z^{n-1}t + \cdots + A_{n-k}A_kz^{n-k}t^k + \cdots + A_0A_nt^n$$
$$+ \cdots$$
$$= A_0 + A_1z + A_1t + A_2z^2 + 2A_2zt + A_2t^2 + \cdots$$
$$+ A_nz^n + A_nnz^{n-1}t + \cdots + A_nn_kz^{n-k}t^k + \cdots + A_nt^n + \cdots,$$

where $$n_k = \frac{(n(n-1)\cdots(n-k+1)}{k!}$$

Equating the coefficients of like powers of z and t in the two members of this equation, we get

$$A_{n-1}A_k \text{ equal always to } A_nn_k.$$

In particular $A_0A_0 = A_0$, therefore $A_0 = 1$. Also

$$A_1A_1 = 2A_2, \quad A_2A_1 = 3A_3,$$
$$A_3A_1 = 4A_4, \cdots, A_{n-1}A_1 = nA_n;$$

or, multiplying these equations together member by member,

$$A_1^n = A_nn!, \text{ or } A_n = \frac{A_1^n}{n!}.$$

A part of the equations among the coefficients are, therefore, sufficient to determine the values of all of them in terms of the one coefficient A_1. But these values will satisfy the remaining equations; for substituting them in the general equation

$$A_{n-k}A_k = A_nn_k,$$

we get $$\frac{A_1^{n-k}}{(n-k)!} \times \frac{A_1^k}{k!} = \frac{A_1^n}{n!} \times \frac{n(n-1)\cdots(n-k+1)}{k!},$$

which is obviously an identical equation.

The coefficient A_1 or, more simply written, A, remains undetermined.

It has been demonstrated, therefore, that to satisfy equation (1), it is only necessary that, $f(z)$ be the sum of an infinite series of the form

$$1 + Az + \frac{A^2}{2!}z^2 + \frac{A^3}{3!}z^3 + \cdots, \tag{2}$$

where A is undetermined; a series which has a sum, i. e. is convergent, for all finite values of z and A. (§ 63, 2, § 66.)

By properly determining A, $f(z)$ may be identified with a^z, for any particular value of a.

If a^z is to be identically equal to the series (2), A must have such a value that

$$a = 1 + A + \frac{A^2}{2!} + \frac{A^3}{3!} + \cdots .$$

Let $$e^z = 1 + z + \frac{z^2}{2!} + \frac{z^3}{3!} + \cdots , \tag{3}$$

where $$e = 1 + 1 + \frac{1}{2!} + \frac{1}{3!} + \cdots ;$$ [2]

Then $$e^A = 1 + A + \frac{A^2}{2!} + \frac{A^3}{3!} + \cdots .$$

Therefore $$a = e^A;$$

or, calling any number which satisfies the equation

$$e^z = a$$

the *logarithm* of a to the base e and writing it $\log_e a$,

$$A = \log_e a.$$

Whence finally,

$$a^z = 1 + (\log_e a)z + \frac{(\log_e a)^2 z^2}{2!} + \frac{(\log_e a)^3 z^3}{3!} + \cdots , \tag{4}$$

a definition of a^z, valid for all finite complex values of a and z, if it may be assumed that $\log_e a$ is a number, whatever the value of a.

The series (3) is commonly called the *exponential series*, and its sum e^z the *exponential function.* It is much more useful than the more general series (2), or (4), because of its greater simplicity; its coefficients do not involve the logarithm, a function not yet fully justified and, as will be shown, to a certain extent indeterminate. Inasmuch, however, as e^z is a particular function of the class a^z, a^z is sometimes called the general exponential function, and series (4) the general exponential series.

74. The Functions Sine and Cosine. It was shown in § 51 that when θ is a real number,

$$e^{i\theta} = \cos\theta + i\sin\theta.$$

But $$\begin{aligned} e^{i\theta} &= 1 + i\theta + \frac{(i\theta)^2}{2!} + \frac{(i\theta)^3}{3!} + \frac{(i\theta)^4}{4!} + \cdots \\ &= 1 - \frac{\theta^2}{2!} + \frac{\theta^4}{4!} - \cdots \\ &\quad + i\left(\theta - \frac{\theta^3}{3!} + \cdots\right). \end{aligned}$$

[2]This number e, the base of the Naperian system of logarithms, is a "transcendental" irrational, transcendental in the sense that there is no algebraic equation with integral coefficients of which it can be a root (see Hermite, Comptes Rendus, LXXVII). π has the same character, as Lindemann proved in 1882, deducing at the same time the first actual demonstration of the impossibility of the famous old problem of squaring the circle by aid of the straight edge and compasses only (see Mathematische Annalen, XX).

Therefore (by § 36, 2, Cor.), for real values of θ

$$\cos\theta = 1 - \frac{\theta^2}{2!} + \frac{\theta^4}{4!} - \cdots, \tag{9.1}$$

and

$$\sin\theta = \theta - \frac{\theta^3}{3!} + \frac{\theta^5}{5!} - \cdots, \tag{9.2}$$

series which both converge for all finite values of θ. Though $\cos\theta$ and $\sin\theta$ only admit of geometrical interpretation when θ is real, it is convenient to continue to use these names for the sums of the series (5) and (6) when θ is complex.

75. Periodicity. When θ is real, evidently neither its sine nor its cosine will be changed if it be increased or diminished by any multiple of four right angles, or 2π; or, if n be any positive integer,

$$\cos(\theta \pm 2n\pi) = \cos\theta,\ \sin(\theta \pm 2n\pi) = \sin\theta,$$

and hence

$$e^{i(\theta\pm 2n\pi)} = e^{i\theta}.$$

The functions $e^{i\theta}$, $\cos\theta$, $\sin\theta$, are on this account called *periodic* functions, with the *modulus of periodicity* 2π.

76. The Logarithmic Function. If $z = e^Z$ and $t = e^T$,

$$zt = e^Z e^T = e^{Z+T}, \qquad \text{§ 73}$$

or

$$\log_e zt = \log_e z + \log_e t. \tag{7}$$

The question again is whether a function exists capable of satisfying this equation, or, more generally, the "functional equation,"

$$f(zt) = f(z) + f(t), \tag{8}$$

for complex values of z and t.

When $z = 0$, (7) becomes

$$\log_e 0 = \log_e 0 + \log_e t,$$

an equation which cannot hold for any value of t for which $\log_e t$ is not zero unless $\log_e 0$ is numerically greater than any finite number whatever. Therefore $\log_e 0$ is infinite.

On the other hand, when $z = 1$, (7) becomes

$$\log_e t = \log_e 1 + \log_e t,$$

so that $\log_e 1$ is zero.

Instead, therefore, of assuming a series with undetermined coefficients for $f(z)$ itself, we assume one for $f(1+z)$, setting

$$f(1+z) = A_1 z + A_2 z^2 + \cdots + A_n z^n + \cdots,$$

and inquire whether the coefficients A_i admit of values which satisfy the functional equation (8) for complex values of z and t.

Now

$$1+z+t=(1+z)\left(1+\frac{t}{1+z}\right), \text{ identically.}$$

$$\therefore f[1+(z+t)]=f(1+z)+f\left(1+\frac{t}{1+z}\right),$$

or

$$\begin{aligned}&A_1(z+t)+A_2(z+t)^2+\cdots+A_n(z+t)^n+\cdots\\=&A_1z+A_2z^2+\cdots+A_nz^n+\cdots\\+&A_1(1+z)^{-1}t+A_2(1+z)^{-2}t^2+\cdots+A_n(1+z)^{-n}t^n+\cdots\end{aligned}$$

Equating the coefficients of the first power of t (§ 72) in the two members of this equation,

$$\begin{aligned}&A_1+2A_2z+3A_3z^2+\cdots+(n+1)A_{n+1}z^n+\cdots\\=&A_1(1-z+z^2-z^3+\cdots+(-1)^nz^n+\cdots);\end{aligned}$$

whence, equating the coefficients of like powers of z,

$$A_1=A_1, 2A_2=-A_1,\cdots,nA_n=(-1)^{n-1}A_1,\cdots,$$

or $$A_2=-\frac{A_1}{2},\cdots,A_n=(-1)^{n-1}\frac{A_1}{n},\cdots.$$

As in the case of the exponential function, a part of the equations among the coefficients are sufficient to determine them all in terms of the one coefficient A_1. But as in that case (by assuming the truth of the binomial theorem for negative integral values of the exponent) it can be readily shown that these values will satisfy the remaining equations also.

The series $\quad z-\dfrac{z^2}{2}+\dfrac{z^3}{3}-\cdots+(-1)^{n-1}\dfrac{z^n}{n}+\cdots$
converges for all values of z whose moduli are less than 1 (§ 62, 3)

For such values, therefore, the function

$$A\left(z-\frac{z^2}{2}+\cdots+(-1)^{n-1}\frac{z^n}{n}+\cdots\right) \tag{9}$$

satisfies the functional equation

$$f[(1+z)(1+t)]=f(1+z)+f(1+t).$$

And since $\quad z\equiv 1-(1-z)$ and $t\equiv 1-(1-t)$,

the function $\quad -A\left(1-z+\dfrac{(1-z)^2}{2}+\cdots+\dfrac{(1-z)^n}{n}+\cdots\right)$

satisfies this equation when written in the simpler form

$$f(zt)=f(z)+f(t),$$

for values of $1-z$ and $1-t$ whose moduli are both less than 1.

1. *Log$_e$b.* To identify the general function $f(1+z)$ with the particular function $\log_e(1+z)$ it is only necessary to give the undetermined coefficient A the value 1.

For since $\log_e(1+z)$ belongs to the class of functions which satisfy the equation (8),

$$\log_e(1+z) = A\left(z - \frac{z^2}{2} + \cdots\right).$$

Therefore

$$e^{\log_e(1+z)} = e^{A\left(z-\frac{z^2}{2}+\cdots\right)}$$

$$= 1 + A\left(z - \frac{z^2}{2} + \cdots\right) + \frac{1}{2!}A^2\left(z - \frac{z^2}{2} + \cdots\right)^2 + \cdots.$$

But $\quad e^{\log_e(1+z)} = 1 + z.$

Hence

$$1 + z = 1 + A\left(z - \frac{z^2}{2} + \cdots\right) + \frac{1}{2!}A^2\left(z - \frac{z^2}{2} + \cdots\right)^2 + \cdots;$$

or, equating the coefficients of the first power of z, $A = 1$.

The coefficients of the higher powers of z in the right number are then identically 0.

It has thus been demonstrated that $\log_e b$ is a number (real or complex), if when b is written in the form $1+z$, the absolute value of z is less than 1. To prove that it is a number for other than such values of b, let $b = \rho e^{i\theta}$, (§ 51), where ρ, as being the modulus of b, is positive.

$$\text{Then} \qquad \log_e b = \log_e \rho + i\theta,$$

and it only remains to prove that $\log_e \rho$ is a number.

Let ρ be written in the form $e^n - (e^n - \rho)$, where e^n is the first integral power of e greater than ρ.

$$\text{Then since} \qquad e^n - (e^n - \rho) \equiv e^n\left(1 - \frac{e^n - \rho}{e^n}\right),$$

$$\log_e \rho = \log_e e^n + \log_e\left(1 - \frac{e^n - \rho}{e^n}\right)$$

$$= n + \log_e\left(1 - \frac{e^n - \rho}{e^n}\right),$$

and $\log_e\left(1 - \frac{e^n - \rho}{e^n}\right)$ is a number since $\frac{e^n - \rho}{e^n}$ is less than 1.

2. *Log$_a$b.* It having now been fully demonstrated that a^z is a number satisfying the equation $a^Z a^T = a^{Z+T}$ for all finite values of a, Z, T; let $a^Z = z$, $a^T = t$, and call Z the *logarithm of z to the base a*, or $\log_a z$, and in like manner T, $\log_a t$.

$$\text{Then, since} \quad zt = a^Z a^T = a^{z+T},$$

$$\log_a(zt) = \log_a z + \log_a t,$$

or $\log_a z$ belongs, like $\log_e z$, to the class of functions which satisfy the functional equation (8).

Pursuing the method followed in the case of $\log_e b$, it will be found that $\log_a(1+z)$ is equal to the series $A\left(z - \frac{z^2}{2} + \cdots\right)$ when $A = \frac{1}{log_e a}$. This number is called the *modulus* of the system of logarithms of which a is base.

77. Indeterminateness of $\log \mathbf{a}$. Since any complex number a may be thrown into the form $\rho e^{i\theta}$,

$$\log_e a = \log_e \rho + i\theta. \tag{10}$$

This, however, is only one of an infinite series of possible values of $\log_e a$. For, since $e^{i\theta} = e^{i(\theta \pm 2n\pi)}$ (§ 75),

$$\log_e a = \log_e \rho e^{i(\theta \pm 2n\pi)} = \log_e \rho + i(\theta \pm 2n\pi),$$

where n may be any positive integer. $\mathrm{Log}_e a$ is, therefore, to a certain extent indeterminate; a fact which must be carefully regarded in using and studying this function.[3] The value given it in (10), for which $n = 0$, is called its principal value.

When a is a positive real number, $\theta = 0$, so that the principal value of $\log_e a$ is real; on the other hand, when a is a negative real number, $\theta = \pi$, or the principal value of $\log_e a$ is the logarithm of the positive number corresponding to a, plus $i\pi$.

78. Permanence of the Remaining Laws of Exponents. Besides the law $a^z a^t = a^{z+t}$ which led to its definition, the function a^z is subject to the laws:

1. $(a^z)^t = a^{zt}$.
2. $(ab)^z = a^z b^z$.[1]

1. $(a^z)^t = a^{zt}$.

$$\text{For} \quad a^z = \left(e^{\log_e a}\right)^z = 1 + (\log_e a)z + \frac{(\log_e a)^2 z^2}{2!} + \cdots \qquad \S\ 73,\ (4)$$

$$= 1 + z\log_e a + \frac{(z\log_e a)^2}{2!} + \cdots$$

$$= e^{z\log_e a}. \qquad \S\ 73,\ (3)$$

$$\therefore (e^{\log_e a})^z = e^{z\log_e a}, \text{ and } \log_e a^z = z\log_e a.$$

From these results it follows that

---

[3]For instance $\log_e(zt)$ is not equal to $\log_e z + \log_e t$ for arbitrarily chosen values of these logarithms, but to $\log_e z + \log_e t \pm i2n\pi$, where n is some positive integer.

$$
\begin{aligned}
(a^z)^t &= e^{\log_e (a^z)^t} \\
&= e^{t \log_e a^z} \\
&= e^{tz \log_e a} \\
&= a^{zt}.
\end{aligned}
$$

2. $(ab)^z = a^z b^z.$

For
$$
\begin{aligned}
(ab)^z &= e^{\log_e (ab)^z} \\
&= e^{z \log_e ab} \\
&= e^{z \log_e a + z \log_e b} && \S\ 76,\ (7) \\
&= e^{z \log_e a} \cdot e^{z \log_e b} && \S\ 73,\ (1) \\
&= a^z \cdot b^z.
\end{aligned}
$$

79. Permanence of the Remaining Law of Logarithms. In like manner, the function $\log_a z$ is subject not only to the law

$$\log_a (zt) = \log_a z + \log_a t,$$

but also to the law

$$\log_a z^t = t \log_a z.$$

For
$$z = a^{\log_a z},$$

and hence
$$
\begin{aligned}
z^t &= (a^{\log_a z})^t \\
&= a^{t \log_a z}. && \S\ 78,\ 1
\end{aligned}
$$

80. Evolution. Consider three complex numbers ζ, z, Z, connected by the equation $\zeta^Z = z$.

This equation gives rise to three problems, each of which is the inverse of the other two. For Z and ζ may be given and z sought; or ζ and z may be given and Z sought; or, finally, z and Z may be given and ζ sought.

The exponential function is the general solution of the first problem (*involution*), and the logarithmic function of the second.

For the third (*evolution*) the symbol $\sqrt[Z]{z}$ has been devised. This symbol does not represent a new function; for it is defined by the equation $(\sqrt[Z]{z})^Z = z$, an equation which is satisfied by the exponential function $z^{\frac{1}{Z}}$.

Like the logarithmic function, $\sqrt[Z]{z}$ is indeterminate, though not always to the same extent. When Z is a positive integer, $\zeta^Z = z$ is an algebraic equation, and by § 56 has Z roots for any one of which $\sqrt[Z]{z}$ is, by definition, a symbol. From the mere fact that $z = t$, therefore, it cannot be inferred that $\sqrt[Z]{z} = \sqrt[Z]{t}$, but only that one of the values of $\sqrt[Z]{z}$ is equal to one of the values of $\sqrt[Z]{t}$. The same remark, of course, applies to the equivalent symbols $z^{\frac{1}{Z}}$, $t^{\frac{1}{Z}}$.

---

[1] $\dfrac{a^z}{a^t} = a^{z-t}$, which is sometimes included among the fundamental laws to which a^z is subject, follows immediately from $a^z a^t = a^{z+t}$ by the definition of division.

81. Permanence of the Binomial Theorem. By aid of the results just obtained, it may readily be demonstrated that the binomial theorem is valid for general complex as well as for rational values of the exponent.

For b being any complex number whatsoever, and the absolute value of z being supposed less than 1,

$$\begin{aligned}(1+z)^b &= e^{b \log_e (1+z)} \\ &= e^{b\left(z - \frac{z^2}{2} + \cdots\right)} \\ &= 1 + bz + \text{ terms involving higher powers of } z.\end{aligned}$$

Therefore let

$$(1+z)^b = 1 + bz + A_2 z^2 + \cdots + A_n z^n + \cdots. \tag{11}$$

Since, then, $(a+Z)^b = a^b\left(1+\frac{z}{a}\right)^b$, § 78, 2

if $\frac{z}{a}$ be substituted for z in (11), and the equation be multiplied throughout by a^b,

$$(a+z)^b = a^b + ba^{b-1}z + A_2 a^{b-2} z^2 + \cdots + A_n a^{b-n} z^n + \cdots. \tag{12}$$

Starting with the identity

$$(1 + \underline{z+t})^b = (\underline{1+z} + t)^b,$$

developing $(1 + \underline{z+t})^b$ by (11) and $(\underline{1+z} + t)^b$ by (12), equating the coefficients of the first power of t in these developments, multiplying the resultant equation by $1+z$, and equating the coefficients of like powers of z in this product, equations are obtained from which values may be derived for the coefficients A_i identical in form with those occurring in the development for $(1+z)^b$ when b is a positive integer.

It may also be shown that these values of the coefficients satisfy the equations which result from equating the coefficients of higher powers of t.

Part II

HISTORICAL.

10. PRIMITIVE NUMERALS.

82. Gesture Symbols. There is little doubt that primitive counting was done on the fingers, that the earliest numeral symbols were groups of the fingers formed by associating a single finger with each individual thing in the group of things whose number it was desired to represent.

Of course the most immediate method of representing the number of things in a group—and doubtless the method first used—is by the presentation of the things themselves or the recital of their names. But to present the things themselves or to recite their names is not in a proper sense to count them; for either the things or their names represent all the properties of the group and not simply the number of things in it. Counting was first done when a group was used to represent the number of things in some other group; of that group it would represent the number only and, therefore, be a true numeral symbol, which it is the sole object of counting to reach.

Counting ignores all the properties of a group except the distinctness or separateness of the things in it and presupposes whatever intelligence is required consciously or unconsciously to abstract this from its remaining properties. On this account, that group serves best to represent numbers, in which the individual differences of the members are least obtrusive. The naturalness of finger-counting, therefore, lies not only in the accessibility of the fingers, in their being always present to the counter, but in this: that the fingers are so similar in form and function that it is almost easier to ignore than to take account of their differences.

But there is other evidence than its intrinsic probability for the priority of finger-counting over any other. Nearly every system of numeral notation of which we have any knowledge is either quinary, decimal, vigesimal, or a mixture of these;[1] that is to say, expresses numbers which are greater than 5 in terms of 5 and lesser numbers, or makes a similar use of 10 or 20. These systems point to primitive methods of reckoning with the fingers of one hand, the fingers of both hands, all the fingers and toes, respectively.

Finger-counting, furthermore, is universal among uncivilized tribes of the present day, even those not far enough developed to have numeral words beyond 2 or 3 representing higher numbers by holding up the appropriate number of fingers.[2]

83. Spoken Symbols. Numeral words—spoken symbols—would naturally arise much later than gesture symbols. Wherever the origin of such a word can be traced, it is found to be either descriptive of the corresponding finger symbol or—when there is nothing characteristic enough about the finger symbol to suggest a word, as is particularly the case with the smaller numbers—the name of some familiar group of things. Thus in the languages of numerous tribes the numeral 5 is simply the word

[1] Pure quinary and vigesimal systems are rare, if indeed they occur at all. As an example of the former, Tylor (Primitive Culture, I, p. 261) instances a Polynesian number series which runs 1, 2, 3, 4, 5, $5 \cdot 1$, $5 \cdot 2$,...; and as an example of the latter, Cantor (Geschichte der Mathematik, p. 8), following Pott, cites the notation of the Mayas of Yucatan who have special words for 20, 400, 8000, 160,000. The Hebrew notation, like the Indo-Arabic, affords an example of a pure decimal notation. Mixed systems are common. Thus the Roman is mixed decimal and quinary, the Aztec mixed vigesimal and quinary. Speaking generally, the quinary and vigesimal systems are more frequent among the lower races, the decimal among the higher. (Primitive Culture, I, p. 262.)

[2] So, for instance, the aborigines of Victoria and the Bororos of Brazil (Primitive Culture, I, p. 244).

for hand, 10 for both hands, 20 for "an entire man" (hands and feet); while 2 is the word for the eyes, the ears, or wings.[3]

As its original meaning is a distinct encumbrance to such a word in its use as a numeral, it is not surprising that the numeral words of the highly developed languages have been so modified that it is for the most part impossible to trace their origin.

The practice of counting with numeral words probably arose much later than the words themselves. There is an artificial element in this sort of counting which does not appertain to primitive counting[4] (see § 5).

One fact is worth reiterating with reference to both the primitive gesture symbols and word symbols for numbers. There is nothing in either symbol to represent the individual characteristics of the things counted or their arrangement. The use of such symbols, therefore, presupposes a conviction that the number of things in a group does not depend on the character of the things themselves or on their collocation, but solely on their maintaining their separateness and integrity.

84. Written Symbols. The earliest *written* symbols for number would naturally be mere groups of strokes—|, ||, |||, etc. Such symbols have a double advantage over gesture symbols: they can be made permanent, and are capable of indefinite extension—there being, of course, no limit to the numbers of strokes which may be drawn.

---

[3]In the language of the Tamanacs on the Orinoco the word for 5 means "a whole hand," the word for 6, "one of the other hand," and so on up to 9; the word for 10 means "both hands," 11, "one to the foot," and so on up to 14; 15 is "a whole foot," 16, "one to the other foot," and so on up to 19; 20 is "one Indian," 40, "two Indians," etc. Other languages rich in digit numerals are the Cayriri, Tupi, Abipone, and Carib of South America; the Eskimo, Aztec, and Zulu (Primitive Culture, I, p. 247).

"Two" in Chinese is a word meaning "ears," in Thibet "wing," in Hottentot "hand." (Gow, Short History of Greek Mathematics, p. 7.) See also Primitive Culture, I, pp. 252–259.

[4]Were there any reason for supposing that primitive counting was done with numeral words, it would be probable that the ordinals, not the cardinals, were the earliest numerals. For the normal order of the cardinals must have been fully recognized before they could be used in counting.

In this connection, see Kronecker, Ueber den Zahlbegriff; Journal für die reine und angewandte Mathematik, Vol. 101, p. 337. Kronecker goes so far as to declare that he finds in the ordinal numbers the natural point of departure for the development of the number concept.

11. HISTORIC SYSTEMS OF NOTATION.

85. Egyptian and Phœnician. This written symbolism did not assume the complicated character it might have had, had counting with written strokes and not with the fingers been the primitive method. Perhaps the written strokes were employed in connection with counting numbers higher than 10 on the fingers to indicate how often all the fingers had been used; or if each stroke corresponded to an individual in the group counted, they were arranged as they were drawn in groups of 10, so that the number was represented by the number of these complete groups and the strokes in a remaining group of less than 10.

At all events, the decimal idea very early found expression in special symbols for 10, 100, and if need be, of higher powers of 10. Such signs are already at hand in the earliest known writings of the Egyptians and Phoenicians in which numbers are represented by unit strokes and the signs for 10, 100, 1000, 10,000, and even 100,000, each repeated up to 9 times.

86. Greek. In two of the best known notations of antiquity, the old Greek notation—called sometimes the Herodianic, sometimes the Attic—and the Roman, a primitive system of counting on the fingers of a single hand has left its impress in special symbols for 5.

In the Herodianic notation the only symbols—apart from certain abbreviations for products of 5 by the powers of 10—are I, Γ (πέντε, 5), Δ (δέκα, 10), H (ἑκατόν, 100), χ (χίλιοι, 1000), M (μυρίοι, 10,000); all of them, except I, it will be noticed, initial letters of numeral words. This is the only notation, it may be added, found in any Attic inscription of a date before Christ. The later and, for the purposes of arithmetic, much inferior notation, in which the 24 letters of the Greek alphabet with three inserted strange letters represent in order the numbers 1, 2, ... 10, 20, ... 100, 200, ... 900, was apparently first employed in Alexandria early in the 3d century B. C., and probably originated in that city.

87. Roman. The Roman notation is probably of Etruscan origin. It has one very distinctive peculiarity: the subtractive meaning of a symbol of lesser value when it precedes one of greater value, as in IV = 4 and in early inscriptions IIX = 8. In nearly every other known system of notation the principle is recognized that the symbol of lesser value shall follow that of greater value and be added to it.

In this connection it is worth noticing that two of the four fundamental operations of arithmetic—addition and multiplication—are involved in the very use of special symbols for 10 and 100, for the one is but a symbol for the *sum* of 10 units, the other a symbol for 10 sums of 10 units each, or for the *product* 10×10. Indeed, addition is primarily only abbreviated counting; multiplication, abbreviated addition. The representation of a number in terms of tens and units, moreover, involves the expression of the result of a division (by 10) in the number of its tens and the result of a subtraction in the number of its units. It does not follow, of course, that the inventors of the notation had any such notion of its meaning or that these inverse operations are, like addition and multiplication, as old as the symbolism itself. Yet the Etrusco-Roman notation testifies to the very respectable antiquity of one of them, subtraction.

88. Indo-Arabic. Associated thus intimately with the four fundamental operations of arithmetic, the character of the numeral notation determines the simplicity or complexity of all reckonings with numbers. An unusual interest, therefore, attaches to the origin of the beautifully clear and simple notation which we are fortunate enough to possess. What a boon that notation is will be appreciated by one who attempts an exercise in division with the Roman or, worst of all, with the later Greek numerals.

The system of notation in current use to-day may be characterized as the positional decimal system. A number is resolved into the sum:

$$a_n 10^n + a_{n-1} 10^{n-1} + \cdots + a_1 10 + a_0,$$

where 10^n is the highest power of 10 which it contains, and $a_n, a_{n-1}, \ldots a_0$ are all numbers less than 10; and then represented by the mere sequence of numbers $a_n a_{n-1} \cdots a_0$—it being left to the *position* of any number a_i in this sequence to indicate the power of 10 with which it is to be associated. For a system of this sort to be complete—to be capable of representing all numbers unambiguously—a symbol (0), which will indicate the absence of any particular power of 10 from the sum $a_n 10^n + a_{n-1} 10^{n-1} + \cdots + a_1 10 + a_0$, is indispensable. Thus without 0, 101 and 11 must both be written 11. But this symbol at hand, any number may be expressed unambiguously in terms of it and symbols for 1, 2, ... 9.

The positional idea is very old. The ancient Babylonians commonly employed a decimal notation similar to that of the Egyptians; but their astronomers had besides this a very remarkable notation, a *sexagesimal* positional system. In 1854 a brick tablet was found near Senkereh on the Euphrates, certainly older than 1600 B. C., on one face of which is impressed a table of the squares, on the other, a table of the cubes of the numbers from 1 to 60. The squares of 1, 2, ... 7 are written in the ordinary decimal notation, but 8^2, or 64, the first number in the table greater than 60, is written 1, 4 $(1 \times 60 + 4)$; similarly 9^2, and so on to 59^2, which is written 58, 1 $(58 \times 60 + 1)$; while 60^2 is written 1. The same notation is followed in the table of cubes, and on other tablets which have since been found. This is a positional system, and it only lacks a symbol for 0 of being a perfect positional system.

The inventors of the 0-symbol and the modern complete decimal positional system of notation were the Indians, a race of the finest arithmetical gifts.

The earlier Indian notation is decimal but not positional. It has characters for 10, 100, etc., as well as for 1, 2, ... 9, and, on the other hand, no 0.

Most of the Indian characters have been traced back to an old alphabet[1] in use in Northern India 200 B. C. The original of each numeral symbol 4, 5, 6, 7, 8 (?), 9, is the initial letter in this alphabet of the corresponding numeral word (see table on page 89,[2] column 1). The characters first occur as numeral signs in certain inscriptions which are assigned to the 1st and 2d centuries A. D. (column 2 of table). Later they took the forms given in column 3 of the table.

When 0 was invented and the positional notation replaced the old notation cannot be exactly determined. It was certainly later than 400 A. D., and there is no evidence that it was earlier than 500 A. D. The earliest known instance of a date written in the new notation is 738 A. D. By the time that 0 came in, the other characters had developed into the so-called Devanagari numerals (table, column 4), the classical numerals of the Indians.

The perfected Indian system probably passed over to the Arabians in 773 A. D., along with certain astronomical writings. However that may be, it was expounded

[1]Dr. Isaac Taylor, in his book "The Alphabet," names this alphabet the Indo-Bactrian. Its earliest and most important monument is the version of the edicts of King Asoka at Kapur-di-giri. In this inscription, it may be added, numerals are denoted by strokes, as |, ||, |||, ||||, |||||.

[2]Columns 1–5, 7, 8 of the table on page 89 are taken from Taylor's Alphabet, II, p. 266; column 6, from Cantor's Geschichte der Mathematik.

in the early part of the 9th century by Alkhwarizmî, and from that time on spread gradually throughout the Arabian world, the numerals taking different forms in the East and in the West.

Europe in turn derived the system from the Arabians in the 12th century, the "Gobar" numerals (table, column 5) of the Arabians of Spain being the pattern forms of the European numerals (table, column 7). The arithmetic founded on the new system was at first called *algorithm* (after Alkhwarizmî), to distinguish it from the arithmetic of the abacus which it came to replace.

A word must be said with reference to this arithmetic on the abacus. In the primitive abacus, or reckoning table, unit counters were used, and a number represented by the appropriate number of these counters in the appropriate columns of the instrument; *e. g.* 321 by 3 counters in the column of 100's, 2 in the column of 10's, and 1 in the column of units. The Romans employed such an abacus in all but the most elementary reckonings, it was in use in Greece, and is in use to-day in China.

Before the introduction of *algorithm*, however, reckoning on the abacus had been improved by the use in its columns of separate characters (called *apices*) for each of the numbers 1, 2, ..., 9, instead of the primitive unit counters. This improved abacus reckoning was probably invented by Gerbert (Pope Sylvester II.), and certainly used by him at Rheims about 970–980, and became generally known in the following century.

Now these apices are not Roman numerals, but symbols which do not differ greatly from the Gobar numerals and are clearly, like them, of Indian origin. In the absence of positive evidence a great controversy has sprung up among historians of mathematics over the immediate origin of the apices. The only earlier mention of them occurs in a passage of the geometry of Boetius, which, if genuine, was written about 500 A. D. Basing his argument on this passage, the historian Cantor urges that the earlier Indian numerals found their way to Alexandria before her intercourse with the East was broken off, that is, before the end of the 4th century, and were transformed by Boetius into the apices. On the other hand, the passage in Boetius is quite generally believed to be spurious, and it is maintained that Gerbert got his apices directly or indirectly from the Arabians of Spain, not taking the 0, either because he did not learn of it, or because, being an abacist, he did not appreciate its value.

At all events, it is certain that the Indo-Arabic numerals, 1, 2, ...9 (not 0), appeared in Christian Europe more than a century before the complete positional system and *algorithm*.

The Indians are the inventors not only of the positional decimal system itself, but of most of the processes involved in elementary reckoning with the system. Addition and subtraction they performed quite as they are performed nowadays; multiplication they effected in many ways, ours among them, but division cumbrously.

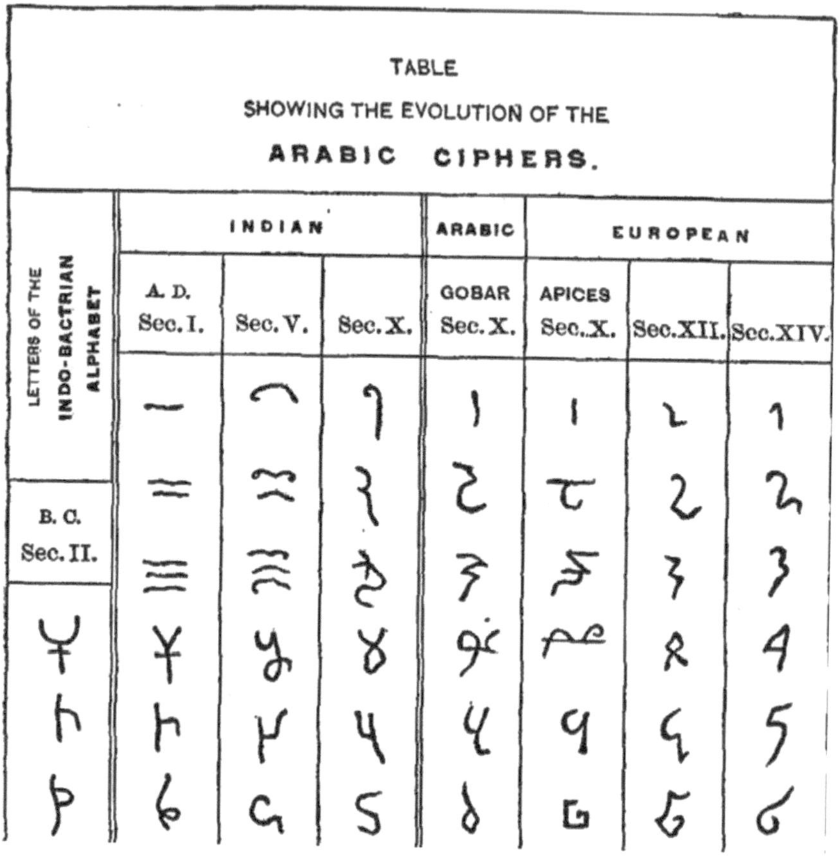
TABLE
SHOWING THE EVOLUTION OF THE
ARABIC CIPHERS.
LETTERS OF THE INDO-BACTRIAN ALPHABET
B. C.
Sec. II.
INDIAN
A. D.
Sec. I.
Sec. V.
Sec. X.
ARABIC
GOBAR
Sec. X.
EUROPEAN
APICES
Sec. X.
Sec. XII.
Sec. XIV.

12. THE FRACTION.

89. Primitive Fractions. Of the artificial forms of number—as we may call the fraction, the irrational, the negative, and the imaginary in contradistinction to the positive integer—all but the fraction are creations of the mathematicians. They were devised to meet purely mathematical rather than practical needs. The fraction, on the other hand, is already present in the oldest numerical records—those of Egypt and Babylonia—was reckoned with by the Romans, who were no mathematicians, and by Greek merchants long before Greek mathematicians would tolerate it in arithmetic.

The primitive fraction was a concrete thing, merely an aliquot part of some larger thing. When a unit of measure was found too large for certain uses, it was subdivided, and one of these subdivisions, generally with a name of its own, made a new unit. Thus there arose fractional units of measure, and in like manner fractional coins.

In time the relation of the sub-unit to the corresponding principal unit came to be abstracted with greater or less completeness from the particular kind of things to which the units belonged, and was recognized when existing between things of other kinds. The relation was generalized, and a pure numerical expression found for it.

90. Roman Fractions. Sometimes, however, the relation was never completely enough separated from the sub-units in which it was first recognized to be generalized. The Romans, for instance, never got beyond expressing all their fractions in terms of the *uncia*, *sicilicus*, etc., names originally of subdivisions of the old unit coin, the *as*.

91. Egyptian Fractions. Races of better mathematical endowments than the Romans, however, had sufficient appreciation of the fractional relation to generalize it and give it an arithmetical symbolism.

The ancient Egyptians had a very complete symbolism of this sort. They represented any fraction whose numerator is 1 by the denominator simply, written as an integer with a dot over it, and resolved all other fractions into sums of such unit fractions. The oldest mathematical treatise known,—a papyrus[1] roll entitled "Directions for Attaining to the Knowledge of All Dark Things," written by a scribe named Ahmes in the reign of Ra-ä-us (therefore before 1700 B. C.), after the model, as he says, of a more ancient work,—opens with a table which expresses in this manner the quotient of 2 by each odd number from 5 to 99. Thus the quotient of 2 by 5 is written $\dot{3}\ \dot{15}$, by which is meant $\frac{1}{3}+\frac{1}{15}$; and the quotient of 2 by 13, $\dot{8}\ \dot{52}\ \dot{104}$. Only $\frac{2}{3}$, among the fractions having numerators which differ from 1, gets recognition as a distinct fraction and receives a symbol of its own.

92. Babylonian or Sexagesimal Fractions. The fractional notation of the Babylonian astronomers is of great interest intrinsically and historically. Like their notation of integers it is a sexagesimal positional notation. The denominator is always 60 or some power of 60 indicated by the position of the numerator, which alone is written. The fraction $\frac{3}{8}$, for instance, which is equal to $\frac{22}{60}+\frac{30}{60^2}$, would in this notation be written 22 30. Thus the ability to represent fractions by a single integer or a sequence of integers, which the Egyptians secured by the use of fractions having a common numerator, 1, the Babylonians found in fractions having common denominators and the principle of position. The Egyptian system is superior in that it gives an exact expression of every quotient, which the Babylonian can in general do only approximately. As regards practical usefulness, however, the Babylonian is beyond

[1] The Rhind papyrus of the British Museum; translated by A. Eisenlohr, Leipzig, 1877.

comparison the better system. Supply the 0-symbol and substitute 10 for 60, and this notation becomes that of the modern decimal fraction, in whose distinctive merits it thus shares.

As in their origin, so also in their subsequent history, the sexagesimal fractions are intimately associated with astronomy. The astronomers of Greece, India, and Arabia all employ them in reckonings of any complexity, in those involving the lengths of lines as well as in those involving the measures of angles. So the Greek astronomer, Ptolemy (150 A. D.), in the *Almagest* (μεγάλη σύνταξις) measures chords as well as arcs in degrees, minutes, and seconds—the degree of chord being the 60th part of the radius as the degree of arc is the 60th part of the arc subtended by a chord equal to the radius.

The sexagesimal fraction held its own as the fraction *par excellence* for scientific computation until the 16th century, when it was displaced by the decimal fraction in all uses except the measurement of angles.

93. Greek Fractions. Fractions occur in Greek writings—both mathematical and non-mathematical—much earlier than Ptolemy, but not in arithmetic.[2] The Greeks drew as sharp a distinction between pure arithmetic, ἀριθμητική, and the art of reckoning, λογιστική, as between pure and metrical geometry. The fraction was relegated to λογιστική. There is no place in a pure science for artificial concepts, no place, therefore, for the fraction in ἀριθμητική; such was the Greek position. Thus, while the metrical geometers—as Archimedes (250 B. C.), in his "Measure of the Circle" (κύκλου μέτρησις), and Hero (120 B. C.)—employ fractions, neither of the treatises on Greek arithmetic before Diophantus (300 A. D.) which have come down to us—the 7th, 8th, 9th books of Euclid's "Elements" (300 B. C.), and the "Introduction to Arithmetic" ([εισαγωγή αριθμητικὴ]) of Nicomachus (100 A. D.)—recognizes the fraction. They do, it is true, recognize the fractional relation. Euclid, for instance, expressly declares that any number is either a multiple, a part, or parts ([μὲρη]), *i. e.* multiple of a part, of every other number (Euc. VII, 4), and he demonstrates such theorems as these:

If A be the same parts of B that C is of D, then the sum or difference of A and C is the same parts of the sum or difference of B and D that A is of B (VII, 6 and 8).

If A be the same parts of B that C is of D, then, alternately, A is the same parts of C that B is of D (VII, 10).

But the relation is expressed by two integers, that which indicates the part and that which indicates the multiple. It is a ratio, and Euclid has no more thought of expressing it except by *two* numbers than he has of expressing the ratio of two geometric magnitudes except by two magnitudes. There is no conception of a single number, the fraction proper, the quotient of one of these integers by the other.

In the αριθμητικὰ of Diophantus, on the other hand, the last and transcendently the greatest achievement of the Greeks in the science of number, the fraction is granted the position in elementary arithmetic which it has held ever since.

[2]The usual method of expressing fractions was to write the numerator with an accent, and after it the denominator twice with a double accent: *e. g.* $\iota\zeta'\ \kappa\alpha''\ \kappa\alpha'' = \frac{17}{21}$. Before sexagesimal fractions came into vogue actual reckonings with fractions were effected by unit fractions, of which only the denominators (doubly accented) were written.

13. ORIGIN OF THE IRRATIONAL.

94. The Discovery of Irrational Lines. The Greeks attributed the discovery of the Irrational to the mathematician and philosopher Pythagoras[1] (525 B. C.).

If, as is altogether probable,[2] the most famous theorem of Pythagoras—that *the square on the hypothenuse of a right triangle is equal to the sum of the squares on the other two sides*—was suggested to him by the fact that $3^2 + 4^2 = 5^2$, in connection with the fact that the triangle whose sides are 3, 4, 5, is right-angled,—for both almost certainly fell within the knowledge of the Egyptians,—he would naturally have sought, after he had succeeded in demonstrating the geometric theorem generally, for number triplets corresponding to the sides of any right triangle as do 3, 4, 5 to the sides of the particular triangle.

The search of course proved fruitless, fruitless even in the case which is geometrically the simplest, that of the isosceles right triangle. To discover that it was *necessarily* fruitless; in the face of preconceived ideas and the apparent testimony of the senses, to conceive that lines may exist which have no common unit of measure, however small that unit be taken; to demonstrate that the hypothenuse and side of the isosceles right triangle actually are such a pair of lines, was the great achievement of Pythagoras.[3]

---

[1]This is the explicit declaration of the most reliable document extant on the history of geometry before Euclid, a chronicle of the ancient geometers which Proclus (A. D. 450) gives in his commentary on Euclid, deriving it from a history written by Eudemus about 330 B. C. This chronicle credits the Egyptians with the discovery of geometry and Thales (600 B. C.) with having first introduced this study into Greece.

Thales and Pythagoras are the founders of the Greek mathematics. But while Thales should doubtless be credited with the first conception of an abstract deductive geometry in contradistinction to the practical empirical geometry of Egypt, the glory of realizing this conception belongs chiefly to Pythagoras and his disciples in the Greek cities of Italy (Magna Græcia); for they established the principal theorems respecting rectilineal figures. To the Pythagoreans the discovery of many of the elementary properties of numbers is due, as well as the geometric form which characterized the Greek theory of numbers throughout its history.

In the middle of the fifth century before Christ Athens became the principal centre of mathematical activity. There Hippocrates of Chios (430 B. C.) made his contributions to the geometry of the circle, Plato (380 B. C.) to geometric method, Theætetus (380 B. C.) to the doctrine of incommensurable magnitudes, and Eudoxus (360 B. C.) to the theory of proportion. There also was begun the study of the conics.

About 300 B. C. the mathematical centre of the Greeks shifted to Alexandria, where it remained.

The third century before Christ is the most brilliant period in Greek mathematics. At its beginning—in Alexandria—Euclid lived and taught and wrote his Elements, collecting, systematizing, and perfecting the work of his predecessors. Later (about 250) Archimedes of Syracuse flourished, the greatest mathematician of antiquity and founder of the science of mechanics; and later still (about 230) Apollonius of Perga, "the great geometer," whose Conics marks the culmination of Greek geometry.

Of the later Greek mathematicians, besides Hero and Diophantus, of whom an account is given in the text, and the great summarizer of the ancient mathematics, Pappus (300 A. D.), only the famous astronomers Hipparchus (130 B. C.) and Ptolemy (150 A. D.) call for mention here. To them belongs the invention of trigonometry and the first trigonometric tables, tables of chords.

The dates in this summary are from Gow's Hist. of Greek Math.

[2]Compare Cantor, Geschichte der Mathematik, p. 153.

[3]His demonstration may easily have been the following, which was old enough in Aristotle's time (340 B. C.) to be made the subject of a popular reference, and which is to be found at the end of the 10th book in all old editions of Euclid's Elements:

95. Consequences of this Discovery in Greek Mathematics. One must know the antecedents and follow the consequences of this discovery to realize its great significance. It was the first recognition of the fundamental difference between the geometric magnitudes and number, which Aristotle formulated brilliantly 200 years later in his famous distinction between the continuous and the discrete, and as such was potent in bringing about that complete banishment of numerical reckoning from geometry which is so characteristic of this department of Greek mathematics in its best, its creative period.

No one before Pythagoras had questioned the possibility of expressing all size relations among lines and surfaces in terms of number,—rational number of course. Indeed, except that it recorded a few facts regarding congruence of figures gathered by observation, the Egyptian geometry was nothing else than a meagre collection of formulas for computing areas. The earliest geometry was metrical.

But to the severely logical Greek no alternative seemed possible, when once it was known that lines exist whose lengths—whatever unit be chosen for measuring them—cannot both be integers, than to have done with number and measurement in geometry altogether. Congruence became not only the final but the sole test of equality. For the study of size relations among unequal magnitudes a pure geometric theory of proportion was created, in which proportion, not ratio, was the primary idea, the method of exhaustions making the theory available for figures bounded by curved lines and surfaces.

The outcome was the system of geometry which Euclid expounds in his Elements and of which Apollonius makes splendid use in his Conics, a system absolutely free from extraneous concepts or methods, yet, within its limits, of great power.

It need hardly be added that it never occurred to the Greeks to meet the difficulty which Pythagoras' discovery had brought to light by inventing an *irrational number*, itself incommensurable with rational numbers. For artificial concepts such as that they had neither talent nor liking.

On the other hand, they did develop the theory of irrational magnitudes as a department of their geometry, the irrational line, surface, or solid being one incommensurable with some chosen (rational) line, surface, solid. Such a theory forms the content of the most elaborate book of Euclid's Elements, the 10th.

96. Approximate Values of Irrationals. In the practical or metrical geometry which grew up after the pure geometry had reached its culmination, and which attained in the works of Hero the Surveyor almost the proportions of our modern elementary mensuration,[4] *approximate values* of irrational numbers played a very important rôle. Nor do such approximations appear for the first time in Hero. In Archimedes' "Measure of the Circle" a number of excellent approximations occur, among them the famous

---

If there be any line which the side and diagonal of a square both contain an exact number of times, let their lengths in terms of this line be a and b respectively; then $b^2 = 2a^2$.

The numbers a and b may have a common factor, γ; so that $a = \alpha\gamma$ and $b = \beta\gamma$, where α and β are prime to each other. The equation $b^2 = 2a^2$ then reduces, on the removal of the factor γ^2 common to both its members, to $\beta^2 = 2\alpha^2$.

From this equation it follows that β^2, and therefore β, is an even number, and hence that α which is prime to β is odd.

But set $\beta = 2\beta'$, where β' is integral, in the equation $\beta^2 = 2\alpha^2$; it becomes $4\beta'^2 = 2\alpha^2$, or $2\beta'^2 = \alpha^2$, whence α^2, and therefore α, is even.

α has thus been proven to be both odd and even, and is therefore not a number.

[4]The formula $\sqrt{s(s-a)(s-b)(s-c)}$ for the area of a triangle in terms of its sides is due to Hero.

approximation $\frac{22}{7}$ for π, the ratio of the circumference of a circle to its diameter. The approximation $\frac{7}{5}$ for $\sqrt{2}$ is reputed to be as old as Plato.

It is not certain how these approximations were effected.[5] They involve the use of some method for extracting square roots. The earliest explicit statement of the method in common use to-day for extracting square roots of numbers (whether exactly or approximately) occurs in the commentary of Theon of Alexandria (380 A. D.) on Ptolemy's *Almagest.* Theon, who like Ptolemy employs sexagesimal fractions, thus finds the length of the side of a square containing $4500°$ to be $67°1'55''$.

97. The Later History of the Irrational is deferred to the chapters which follow (§§ 106, 108, 112, 121, 129).

It will be found that the Indians permitted the simplest forms of irrational numbers, surds, in their algebra, and that they were followed in this by the Arabians and the mathematicians of the Renaissance, but that the general irrational did not make its way into algebra until after Descartes.

[5] Many attempts have been made to discover the methods of approximation used by Archimedes and Hero from an examination of their results, but with little success. The formula $\sqrt{a^2 \pm b} = a \pm \frac{b}{2a}$ will account for some of the simpler approximations, but no single method or set of methods have been found which will account for the more difficult. See Günther: Die quadratischen Irrationalitäten der Alten und deren Entwicklungsmethoden. Leipzig, 1882. Also in Handbuch der klassischen Altertums-Wissenschaft, 11ter. Halbband.

14. ORIGIN OF THE NEGATIVE AND THE IMAGINARY. THE EQUATION.

98. The Equation in Egyptian Mathematics. While the irrational originated in geometry, the negative and the imaginary are of purely algebraic origin. They sprang directly from the algebraic equation.

The authentic history of the equation, like that of geometry and arithmetic, begins in the book of the old Egyptian scribe Ahmes. For Ahmes, quite after the present method, solves numerical problems which admit of statement in an equation of the first degree involving one unknown quantity.[1]

99. In the Earlier Greek Mathematics. The equation was slow in arousing the interest of Greek mathematicians. They were absorbed in geometry, in a geometry whose methods were essentially non-algebraic.

To be sure, there are occasional signs of a concealed algebra under the closely drawn geometric cloak. Euclid solves three geometric problems which, stated algebraically, are but the three forms of the quadratic; $x^2 + ax = b^2$, $x^2 = ax + b^2$, $x^2 + b^2 = ax$.[2] And the Conics of Apollonius, so astonishing if regarded as a product of the pure geometric method used in its demonstrations, when stated in the language of algebra, as recently it has been stated by Zeuthen,[3] almost convicts its author of the use of algebra as his instrument of investigation.

100. Hero. But in the writings of Hero of Alexandria (120 B. C.) the equation first comes clearly into the light again. Hero was a man of practical genius whose aim was to make the rich pure geometry of his predecessors available for the surveyor. With him the rigor of the old geometric method is relaxed; proportions, even equations, among the *measures* of magnitudes are permitted where the earlier geometers allow only proportions among the magnitudes themselves; the theorems of geometry are stated metrically, in formulas; and more than all this, the equation becomes a recognized geometric instrument.

Hero gives for the diameter of a circle in terms of s, the sum of diameter, circumference, and area, the formula:[4]

$$d = \frac{\sqrt{154s + 841} - 29}{11}$$

He could have reached this formula only by *solving a quadratic equation*, and that not geometrically,—the nature of the oddly constituted quantity s precludes that supposition,—but by a purely algebraic reckoning like the following:

The area of a circle in terms of its diameter being $\frac{\pi d^2}{4}$, the length of its circumference πd, and π according to Archimedes' approximation $\frac{22}{7}$, we have the equation:

$$s = d + \frac{\pi d^2}{4} + \pi d, \quad \text{or} \quad \frac{11}{14}d^2 + \frac{29}{7}d = s.$$

Clearing of fractions, multiplying by 11, and completing the square,

$$121d^2 + 638d + 841 = 154s + 841,$$

[1] His symbol for the unknown quantity is the word *hau*, meaning heap.

[2] Elements, VI, 29, 28; Data, 84, 85.

[3] Die Lehre von den Kegelschnitten im Altertum. Copenhagen, 1886.

[4] See Cantor; Geschichte der Mathematik, p. 341.

whence

$$11d + 29 = \sqrt{154s + 841},$$

or

$$d = \frac{\sqrt{154s + 841} - 29}{11}.$$

Except that he lacked an algebraic symbolism, therefore, Hero was an algebraist, an algebraist of power enough to solve an affected quadratic equation.

101. Diophantus. (300 A. D.?). The last of the Greek mathematicians, Diophantus of Alexandria, was a great algebraist.

The period between him and Hero was not rich in creative mathematicians, but it must have witnessed a gradual development of algebraic ideas and of an algebraic symbolism.

At all events, in the ἀριθμητικά of Diophantus the algebraic equation has been supplied with a symbol for the unknown quantity, its powers and the powers of its reciprocal to the 6th, and a symbol for equality. Addition is represented by mere juxtaposition, but there is a special symbol, see Figure A, for subtraction. On the other hand, there are no general symbols for known quantities,—symbols to serve the purpose which the first letters of the alphabet are made to serve in elementary algebra nowadays,—therefore no literal coefficients and no general formulas.

FIG. A.

With the symbolism had grown up many of the formal rules of algebraic reckoning also. Diophantus prefaces the αριθμητικὰ with rules for the addition, subtraction, and multiplication of polynomials. He states expressly that the product of two subtractive terms is additive.

The αριθμητικὰ itself is a collection of problems concerning numbers, some of which are solved by determinate algebraic equations, some by indeterminate.

Determinate equations are solved which have given positive integers as coefficients, and are of any of the forms $ax^m = bx^n$, $ax^2 + bx = c$, $ax^2 + c = bx$, $ax^2 = bx + c$; also a single cubic equation, $ax^3 + x = 4x^2 + 4$. In reducing equations to these forms, equal quantities in opposite members are cancelled and subtractive terms in either member are rendered additive by transposition to the other member.

The indeterminate equations are of the form $y^2 = ax^2 + bx + c$, Diophantus regarding any pair of positive *rational* numbers (integers or fractions) as a solution which, substituted for y and x, satisfy the equation.[5] These equations are handled with marvellous dexterity in the αριθμητικὰ. No effort is made to develop general comprehensive methods, but each exercise is solved by some clever device suggested by its individual peculiarities. Moreover, the discussion is never exhaustive, one solution sufficing when the possible number is infinite. Yet until some trace of indeterminate equations earlier than the αριθμητικὰ is discovered, Diophantus must rank as the originator of this department of mathematics.

The determinate quadratic is solved by the method which we have already seen used by Hero. The equation is first multiplied throughout by a number which renders

[5]The designation "Diophantine equations," commonly applied to indeterminate equations of the first degree when investigated for integral solutions, is a striking misnomer. Diophantus nowhere considers such equations, and, on the other hand, allows fractional solutions of indeterminate equations of the second degree.

the coefficient of x^2 a perfect square, the "square is completed," the square root of both members of the equation taken, and the value of x reckoned out from the result. Thus from $ax^2 + c = bx$ is derived first the equation

$$a^2x^2 + ac = abx,$$

$$\text{then}\quad a^2x^2 - abx + \left(\frac{b}{2}\right)^2 = \left(\frac{b}{2}\right)^2 - ac,$$

$$\text{then}\quad ax - \frac{b}{2} = \sqrt{\left(\frac{b}{2}\right)^2 - ac},$$

$$\text{and finally,}\quad x = \frac{\frac{b}{2} + \sqrt{\left(\frac{b}{2}\right)^2 - ac}}{a}.$$

The solution is regarded as possible only when the number under the radical is a perfect square (it must, of course, be positive), and only one root—that belonging to the positive value of the radical—is ever recognized.

Thus the number system of Diophantus contained only the positive integer and fraction; the irrational is excluded; and as for the negative, there is no evidence that a Greek mathematician ever conceived of such a thing,—certainly not Diophantus with his three classes and one root of affected quadratics. The position of Diophantus is the more interesting in that in the *αριθμητικὰ* the Greek science of number culminates.

102. The Indian Mathematics. The pre-eminence in mathematics passed from the Greeks to the Indians. Three mathematicians of India stand out above the rest: *Âryabhaṭṭa* (born 476 A. D.), *Brahmagupta* (born 598 A. D.), *Bhâskara* (born 1114 A. D.) While all are in the first instance astronomers, their treatises also contain full expositions of the mathematics auxiliary to astronomy, their reckoning, algebra, geometry, and trigonometry.[6]

An examination of the writings of these mathematicians and of the remaining mathematical literature of India leaves little room for doubt that the Indian geometry was taken bodily from Hero, and the algebra—whatever there may have been of it before Âryabhaṭṭa—at least powerfully affected by Diophantus. Nor is there occasion for surprise in this. Âryabhaṭṭa lived two centuries after Diophantus and six after Hero, and during those centuries the East had frequent communication with the West through various channels. In particular, from Trajan's reign till later than 300 A. D. an active commerce was kept up between India and the east coast of Egypt by way of the Indian Ocean.

Greek geometry and Greek algebra met very different fates in India. The Indians lacked the endowments of the geometer. So far from enriching the science with new discoveries, they seem with difficulty to have kept alive even a proper understanding of Hero's metrical formulas. But algebra flourished among them wonderfully. Here the fine talent for reckoning which could create a perfect numeral notation, supported by a talent equally fine for symbolical reasoning, found a great opportunity and made great achievements. With Diophantus algebra is no more than an art by which disconnected numerical problems are solved; in India it rises to the dignity of a science, with general methods and concepts of its own.

[6]The mathematical chapters of Brahmagupta and Bhâskara have been translated into English by Colebrooke: "Algebra, Arithmetic, and Mensuration, from the Sanscrit of Brahmagupta and Bhâskara," 1817; those of Âryabhaṭṭa into French by L. Rodet (Journal Asiatique, 1879).

103. Its Algebraic Symbolism. First of all, the Indians devised a complete, and in most respects adequate, symbolism. Addition was represented, as by Diophantus, by mere juxtaposition; subtraction, exactly as addition, except that a dot was written over the coefficient of the subtrahend. The syllable *bha* written after the factors indicated a product; the divisor written under the dividend, a quotient; a syllable, *ka*, written before a number, its (irrational) square root; one member of an equation placed over the other, their equality. The equation was also provided with symbols for any number of unknown quantities and their powers.

104. Its Invention of the Negative. The most note-worthy feature of this symbolism is its representation of subtraction. To remove the subtractive symbol from between minuend and subtrahend (where Diophantus had placed his symbol, see Figure A.) to attach it wholly to the subtrahend and then connect this modified subtrahend with the minuend additively, is, formally considered, to transform the subtraction of a positive quantity into the addition of the corresponding negative. It suggests what other evidence makes certain, that *algebra owes to India the immensely useful concept of the absolute negative.*

Thus one of these dotted numbers is allowed to stand by itself as a member of an equation. Bhâskara recognizes the double sign of the square root, as well as the impossibility of the square root of a negative number (which is very interesting, as being the first dictum regarding the imaginary), and no longer ignores either root of the quadratic. More than this, recourse is had to the same expedients for interpreting the negative, for attaching a concrete physical idea to it, as are in common use to-day. The primary meaning of the very name given the negative was *debt*, as that given the positive was *means*. The opposition between the two was also pictured by lines described in opposite directions.

105. Its Use of Zero. But the contributions of the Indians to the fund of algebraic concepts did not stop with the absolute negative.

They made a number of 0, and though some of their reckonings with it are childish, Bhâskara, at least, had sufficient understanding of the nature of the "quotient" $\frac{a}{0}$ (infinity) to say "it suffers no change, however much it is increased or diminished." He associates it with Deity.

106. Its Use of Irrational Numbers. Again, the Indians were the first to reckon with irrational square roots as with numbers; Bhâskara extracting square roots of binomial surds and rationalizing irrational denominators of fractions even when these are polynomial. Of course they were as little able rigorously to justify such a procedure as the Greeks; less able, in fact, since they had no equivalent of the method of exhaustions. But it probably never occurred to them that justification was necessary; they seem to have been unconscious of the gulf fixed between the discrete and continuous. And here, as in the case of 0 and the negative, with the confidence of apt and successful reckoners, they were ready to pass immediately from numerical to purely symbolical reasoning, ready to trust their processes even where formal demonstration of the right to apply them ceased to be attainable. Their skill was too great, their instinct too true, to allow them to go far wrong.

107. Determinate and Indeterminate Equations in Indian Algebra. As regards equations—the only changes which the Indian algebraists made in the treatment of determinate equations were such as grew out of the use of the negative. This brought the triple classification of the quadratic to an end and secured recognition for both roots of the quadratic.

Brahmagupta solves the quadratic by the rule of Hero and Diophantus, of which he gives an explicit and general statement. Çrîdhara, a mathematician of some distinction belonging to the period between Brahmagupta and Bhâskara, made the improvement of this method which consists in first multiplying the equation throughout by four times the coefficient of the square of the unknown quantity and so preventing the occurrence of fractions under the radical sign.[7]

Bhâskara also solves a few cubic and biquadratic equations by special devices.

The theory of indeterminate equations, on the other hand, made great progress in India. The achievements of the Indian mathematicians in this beautiful but difficult department of the science are as brilliant as those of the Greeks in geometry. They created the doctrine of the indeterminate equation of the first degree, $ax + by = c$, which they treated for integral solutions by the method of continued fractions in use today. They worked also with equations of the second degree of the forms $ax^2 + b = cy^2$, $xy = ax + by + c$, originating general and comprehensive methods where Diophantus had been content with clever devices.

108. The Arabian Mathematics. The Arabians were the instructors of modern Europe in the ancient mathematics. The service which they rendered in the case of the numeral notation and reckoning of India they rendered also in the case of the geometry, algebra, and astronomy of the Greeks and Indians. Their own contributions to mathematics are unimportant. Their receptiveness for mathematical ideas was extraordinary, but they had little originality.

The history of Arabian mathematics begins with the reign of Almanṣûr (754–775),[8] the second of the Abbasid caliphs.

It is related (by Ibn-al-Adamî, about 900) that in this reign, in the year 773, an Indian brought to Bagdad certain astronomical writings of his country, which contained a method called "Sindhind," for computing the motions of the stars,—probably portions of the Siddhânta of Brahmagupta,—and that Alfazârî was commissioned by the caliph to translate them into Arabic.[9] Inasmuch as the Indian astronomers put full expositions of their reckoning, algebra, and geometry into their treatises, Alfazârî's translation laid open to his countrymen a rich treasure of mathematical ideas and methods.

It is impossible to set a date to the entrance of Greek ideas. They must have made themselves felt at Damascus, the residence of the later Omayyad caliphs, for that city had numerous inhabitants of Greek origin and culture. But the first translations of Greek mathematical writings were made in the reign of Hârûn Arraschîd (786–809), when Euclid's Elements and Ptolemy's Almagest were put into Arabic. Later on, translations were made of Archimedes, Apollonius, Hero, and last of all, of Diophantus (by Abû'l Wafâ, 940–998).

The earliest mathematical author of the Arabians is Alkhwarizmî, who flourished in the first quarter of the 9th century. Besides astronomical tables, he wrote a treatise on algebra and one on reckoning (elementary arithmetic). The latter has already been mentioned. It is an exposition of the positional reckoning of India, the reckoning which mediæval Europe named after him *Algorithm*.

---

[7]This method still goes under the name "Hindoo method."

[8]It was Almanṣûr who transferred the throne of the caliphs from Damascus to Bagdad which immediately became not only the capital city of Islam, but its commercial and intellectual centre.

[9]This translation remained the guide of the Arabian astronomers until the reign of Almamûn (813–833), for whom Alkhwarizmî prepared his famous astronomical tables (820). Even these were based chiefly on the "Sindhind," though some of the determinations were made by methods of the Persians and Ptolemy.

The treatise on algebra bears a title in which the word *Algebra* appears for the first time: viz., *Aldjebr walmukâbala*. Aldjebr (*i. e.* reduction) signifies the making of all terms of an equation positive by transferring negative terms to the opposite member of the equation; *almukâbala* (*i. e.* opposition), the cancelling of equal terms in opposite members of an equation.

Alkhwarizmî's classification of equations of the 1st and 2d degrees is that to which these processes would naturally lead, viz.:

$$\begin{array}{lll} ax^2 = bx, & bx^2 = c, & bx = c, \\ x^2 + bx = c, & x^2 + c = bx, & x^2 = bx + c. \end{array}$$

These equations he solves separately, following up the solution in each case with a geometric demonstration of its correctness. He recognizes both roots of the quadratic when they are positive. In this respect he is Indian; in all others—the avoidance of negatives, the use of geometric demonstration—he is Greek.

Besides Alkhwarizmî, the most famous algebraists of the Arabians were *Alkarchî* and *Alchayyâmî*, both of whom lived in the 11th century.

Alkarchî gave the solution of equations of the forms:

$$ax^{2p} + bx^p = c, ax^{2p} + c = bx^p, bx^p + c = ax^{2p}.$$

He also reckoned with irrationals, the equations

$$\sqrt{8} + \sqrt{18} = \sqrt{50}, \sqrt[3]{54} - \sqrt[3]{2} = \sqrt[3]{16},$$

being pretty just illustrations of his success in this field.

Alchayyâmî was the first mathematician to make a systematic investigation of the cubic equation. He classified the various forms which this equation takes when all its terms are positive, and solved each form geometrically—by the intersections of conics.[10] A pure algebraic solution of the cubic he believed impossible.

Like Alkhwarizmî, Alkarchî and Alchayyâmî were Eastern Arabians. But early in the 8th century the Arabians conquered a great part of Spain. An Arabian realm was established there which became independent of the Bagdad caliphate in 747, and endured for 300 years. The intercourse of these Western Arabians with the East was not frequent enough to exercise a controlling influence on their æsthetic or scientific development. Their mathematical productions are of a later date than those of the East and almost exclusively arithmetico-algebraic. They constructed a formal algebraic notation which went over into the Latin translations of their writings and rendered the path of the Europeans to a knowledge of the doctrine of equations easier than it would have been, had the Arabians of the East been their only instructors. The best known of their mathematicians are *Ibn Aflah* (end of 11th century), *Ibn Albannâ* (end of 13th century), *Alkasâdî* (15th century).

---

[10]Thus suppose the equation $x^3 + bx = a$, given.

For b substitute the quantity p^2, and for a, p^2r. Then $x^3 = p^3(r - x)$.

Now this equation is the result of eliminating y from between the two equations, $x^2 = py$, $y^2 = x(r - x)$; the first of which is the equation of a parabola, the second, of a circle.

Let these two curves be constructed; they will intersect in one real point distinct from the origin, and the abscissa of this point is a root of $x^3 + bx = a$. See Hankel, Geschichte der Mathematik, p. 279.

This method is of greater interest in the history of geometry than in that of algebra. It involves an anticipation of some of the most important ideas of Descartes' *Géométrie* (see p. 118).

109. Arabian Algebra Greek rather than Indian. Thus, of the three greater departments of the Arabian mathematics, the Indian influence gained the mastery in reckoning only.

The Arabian geometry is Greek through and through.

While the algebra contains both elements, the Greek predominates. Indeed, except that both roots of the quadratic are recognized, the doctrine of the determinate equation is altogether Greek. It avoids the negative almost as carefully as Diophantus does; and in its use of the geometric method of demonstration it is actuated by a spirit less modern still—the spirit in which Euclid may have conceived of algebra when he solved his geometric quadratics.

The theory of indeterminate equations seldom goes beyond Diophantus; where it does, it is Indian.

The Arabian trigonometry is based on Ptolemy's, but is its superior in two important particulars. It employs the sine where Ptolemy employs the chord (being in this respect Indian), and has an algebraic instead of a geometric form. Some of the methods of approximation used in reckoning out trigonometric tables show great cleverness. Indeed, the Arabians make some amends for their ill-advised return to geometric algebra by this excellent achievement in algebraic geometry.

The preference of the Arabians for Greek algebra was especially unfortunate in respect to the negative, which was in consequence forced to repeat in Europe the fight for recognition which it had already won in India.

110. Mathematics in Europe before the Twelfth Century. The Arabian mathematics found entrance to Christian Europe in the 12th century. During this century and the first half of the next a good part of its literature was translated into Latin.

Till then the plight of mathematics in Europe had been miserable enough. She had no better representatives than the Romans, the most deficient in the sense for mathematics of all cultured peoples, ancient or modern; no better literature than the collection of writings on surveying known as the *Codex Arcerianus*, and the childish arithmetic and geometry of Boetius.

Prior to the 10th century, however, Northern Europe had not sufficiently emerged from barbarism to call even this paltry mathematics into requisition. What learning there was was confined to the cloisters. Reckoning (*computus*) was needed for the Church calendar and was taught in the cloister schools established by Alcuin (735–804) under the patronage of Charlemagne. Reckoning was commonly done on the fingers. Not even was the multiplication table generally learned. Reference would be made to a written copy of it, as nowadays reference is made to a table of logarithms. The Church did not need geometry, and geometry in any proper sense did not exist.

111. Gerbert. But in the 10th century there lived a man of true scientific interests and gifts, Gerbert,[11] Bishop of Rheims, Archbishop of Ravenna, and finally Pope Sylvester II. In him are the first signs of a new life for mathematics. His achievements, it is true, do not extend beyond the revival of Roman mathematics, the authorship of a geometry based on the *Codex Arcerianus*, and a method for effecting division on the abacus with apices. Yet these achievements are enough to place him far above his contemporaries. His influence gave a strong impulse to mathematical studies where interest in them had long been dead. He is the forerunner of the intellectual activity ushered in by the translations from the Arabic, for he brought to life the feeling of the need for mathematics which these translations were made to satisfy.

[11] See §88.

112. Entrance of the Arabian Mathematics. Leonardo. It was the elementary branch of the Arabian mathematics which took root quickest in Christendom—reckoning with nine digits and 0.

Leonardo of Pisa—*Fibonacci*, as he was also called—did great service in the diffusion of the new learning through his *Liber Abaci* (1202 and 1228), a remarkable presentation of the arithmetic and algebra of the Arabians, which remained for centuries the fund from which reckoners and algebraists drew and is indeed the foundation of the modern science.

The four fundamental operations on integers and fractions are taught after the Arabian method; the extraction of the square root and the doctrine of irrationals are presented in their pure algebraic form; quadratic equations are solved and applied to quite complicated problems; *negatives are accepted when they admit of interpretation as debt.*

The last fact illustrates excellently the character of the *Liber Abaci.* It is not a mere translation, but an independent and masterly treatise in one department of the new mathematics.

Besides the *Liber Abaci*, Leonardo wrote the *Practica Geometriae*, which contains much that is best of Euclid, Archimedes, Hero, and the elements of trigonometry; also the *Liber Quadratorum*, a collection of original algebraic problems most skilfully handled.

113. Mathematics during the Age of Scholasticism. Leonardo was a great mathematician,[12] but fine as his work was, it bore no fruit until the end of the 15th century. In him there had been a brilliant response to the Arabian impulse. But the awakening was only momentary; it quickly yielded to the heavy lethargy of the "dark" ages.

The age of scholasticism, the age of devotion to the forms of thought, logic and dialectics, is the age of greatest dulness and confusion in mathematical thinking.[13]

[12]Besides Leonardo there flourished in the first quarter of the 13th century an able German mathematician, *Jordanus Nemorarius.* He was the author of a treatise entitled *De numeris datis*, in which known quantities are for the first time represented by letters, and of one *De trangulis* which is a rich though rather systemless collection of theorems and problems principally of Greek and Arabian origin. See Günther: Geschichte des mathemathischen Unterrichts im deutschen Mittelalter, p. 156.

[13]Compare Hankel, Geschichte der Mathematik, pp. 349–352. To the unfruitfulness of these centuries the *Summa* of *Luca Pacioli* bears witness. This book, which has the distinction of being the earliest book on algebra printed, appeared in 1494, and embodies the arithmetic, algebra, and geometry of the time just preceding the Renaissance. It contains not an idea or method not already presented by Leonardo. Even in respect to algebraic symbolism it surpasses the *Liber Abaci* only to the extent of using abbreviations for a few frequently recurring words, as p. for "plus," and R. for "res" (the unknown quantity). And this is not to be regarded as original with Pacioli for the Arabians of Leonardo's time made a similar use of abbreviations. In a translation made by Gerhard of Cremona (12th century) from an unknown Arabic original the letters r (radix), c (census), d (dragma) are used to represent the unknown quantity, its square, and the absolute term respectively.

The *Summa* of Pacioli has great merits, notwithstanding its lack of originality. It satisfied the mathematical needs of the time. It is very comprehensive, containing full and excellent instruction in the art of reckoning after the methods of Leonardo, for the merchant-man, and a great variety of matter of a purely theoretical interest also—representing the elementary theory of numbers, algebra, geometry, and the application of algebra to geometry. Compare Cantor, Geschichte der Mathematik, II, p. 308.

It should be added that the 15th century produced a mathematician who deserves a distinguished place in the general history of mathematics on account of his contributions to

Algebra owes the entire period but a single contribution; the concept of the fractional power. Its author was Nicole Oresme (died 1382), who also gave a symbol for it and the rules by which reckoning with it is governed.

114. The Renaissance. Solution of the Cubic and Biquadratic Equations. The first achievement in algebra by the mathematicians of the Renaissance was the algebraic solution of the cubic equation: a fine beginning of a new era in the history of the science.

The cubic $x^3 + mx = n$ was solved by *Ferro* of Bologna in 1505, and a second time and independently, in 1535, by Ferro's countryman, *Tartaglia*, who by help of a transformation made his method apply to $x^3 \pm mx^2 = \pm n$ also. But *Cardan* of Milan was the first to publish the solution, in his *Ars Magna*,[14] 1545.

The *Ars Magna* records another brilliant discovery: the solution—after a general method—of the biquadratic $x^4 + 6x^2 + 36 = 60x$ by *Ferrari*, a pupil of Cardan.

Thus in Italy, within fifty years of the new birth of algebra, after a pause of sixteen centuries at the quadratic, the limits of possible attainment in the algebraic solution of equations were reached; for the algebraic solution of the general equation of a degree higher than 4 is impossible, as was first demonstrated by Abel.[15]

The general solution of higher equations proving an obstinate problem, nothing was left the searchers for the roots of equations but to devise a method of working them out approximately. In this the French mathematician *Vieta* (1540–1603) was successful, his method being essentially the same as that now known as Newton's.

115. The Negative in the Algebra of this Period. First Appearance of the Imaginary. But the general equation presented other problems than the discovery of rules for obtaining its roots; the nature of these roots and the relations between them and the coefficients of the equation invited inquiry.

We witness another phase of the struggle of the negative for recognition. The imaginary is now ready to make common cause with it.

Already in the *Ars Magna* Cardan distinguishes between *numeri veri*—the positive integer, fraction, and irrational,—and *numeri ficti*, or *falsi*—the negative and the square root of the negative. Like Leonardo, he tolerates negative roots of equations when they admit of interpretation as "debitum," not otherwise. While he has no thought of accepting imaginary roots, he shows that if $5 + \sqrt{-15}$ be substituted for x in $x(10 - x) = 40$, that equation is satisfied; which, of course, is all that is meant nowadays when $5 + \sqrt{-15}$ is called a root. His declaration that $5 \pm \sqrt{-15}$ are "vere sophistica" does not detract from the significance of this, the earliest recorded instance of reckoning with the imaginary. It ought perhaps to be added that Cardan is not always so successful in these reckonings; for in another place he sets

$$\frac{1}{4}\left(-\sqrt{-\frac{1}{4}}\right) = \sqrt{\frac{1}{64}} = \frac{1}{8}$$

Following Cardan, *Bombelli*[16] reckoned with imaginaries to good purpose, explaining by their aid the irreducible case in Cardan's solution of the cubic.

On the other hand, neither Vieta nor his distinguished follower, the Englishman *Harriot* (1560–1621), accept even negative roots; though Harriot does not hesitate to

trigonometry, the astronomer *Regiomontanus* (1436–1476). Like Jordanus, he was a German.

[14] The proper title of this work is: "Artis magnae sive de regulis Algebraicis liber unus." It has stolen the title of Cardan's "Ars magna Arithmeticae," published at Basel, 1570.

[15] Mémoire sur les Equations Algébriques: Christiania, 1826. Also in Crelle's Journal, I, p. 65.

[16] L'Algebra, 1579. He also formally states rules for reckoning with $\pm\sqrt{-1}$ and $a + b\sqrt{-1}$.

perform algebraic reckonings on negatives, and even allows a negative to constitute one member of an equation.

116. Algebraic Symbolism. Vieta and Harriot. Vieta and Harriot, however, did distinguished service in perfecting the symbolism of algebra; Vieta, by the systematic use of letters to represent known quantities,—algebra first became "literal" or "universal arithmetic" in his hands,[17]—Harriot, by ridding algebraic statements of every non-symbolic element, of everything but the letters which represent quantities known as well as unknown, symbols of operation, and symbols of relation. Harriot's *Artis Analyticae Praxis* (1631) has quite the appearance of a modern algebra.[18]

117. Fundamental Theorem of Algebra. Harriot and Girard. Harriot has been credited with the discovery of the "fundamental theorem" of algebra—the theorem that the number of roots of an algebraic equation is the same as its degree. The *Artis Analyticae Praxis* contains no mention of this theorem—indeed, by ignoring negative and imaginary roots, leaves no place for it; yet Harriot develops systematically a method which, if carried far enough, leads to the discovery of this theorem as well as to the relations holding between the roots of an equation and its coefficients.

By multiplying together binomial factors which involve the unknown quantity, and setting their product equal to 0, he builds "canonical" equations, and shows that the roots of these equations—the only roots, he says—are the positive values of the unknown quantity which render these binomial factors 0. Thus he builds $aa-ba-ca = -bc$, in which a is the unknown quantity, out of the factors $a-b, a+c$, and proves that b is a root of this equation and the only root, the negative root c being totally ignored.

While no attempt is made to show that if the terms of a "common" equation be collected in one member, this can be separated into binomial factors, the case of

[17]There are isolated instances of this use of letters much earlier than Vieta in the *De numeris datis* of Jordanus Nemorarius, and in the *Algorithmus demonstratus* of the same author. But the credit of making it the general practice of algebraists belongs to Vieta.

[18]One has only to reflect how much of the power of algebra is due to its admirable symbolism to appreciate the importance of the *Artis Analyticae Praxis*, in which this symbolism is finally established. But one addition of consequence has since been made to it, integral and fractional exponents introduced by Descartes (1637) and Wallis (1659).

Harriot substituted small letters for the capitals used by Vieta, but followed Vieta in representing known quantities by consonants and unknown by vowels. The present convention of representing known quantities by the earlier letters of the alphabet, unknown by the later, is due to Descartes.

Vieta's notation is unwieldy and ill adapted to purposes of algebraic reckoning. Instead of restricting itself, as Harriot's does, to the use of brief and easily apprehended conventional symbols, it also employs words subject to the rules of syntax. Thus for $A^3 - 3B^2A = Z$ (or $aaa - 3bba = z$, as Harriot would have written it), Vieta writes *A cubus - B quad 3 in A aequatur Z solido*. In this respect Vieta is inferior not only to Harriot, but to several of his predecessors and notably to his contemporary, the Dutch mathematician Stevinus (1548–1620), who would, for instance, have written x^2+3x-8 as $1 * +3 * -8*$. The geometric affiliations of Vieta's notation are obvious. It suggests the Greek arithmetic.

It is surprising that algebraic symbolism should owe so little to the great Italian algebraists of the 16th century. Like Pacioli (see note, p. 113) they were content with a few abbreviations for words, a "syncopated" notation, as it has been called, and an incomplete one at that.

The current symbols of operation and relation are chiefly of English and German origin, having been invented or adopted as follows: viz. $=$, by *Recorde* in 1556; $\surd$, by *Rudolf* in 1525; the *vinculum*, by *Vieta* in 1591; *brackets*, by *Bombelli*, 1572; $\div$, by *Rahn* in 1659; $\times, >, <$, by *Harriot* in 1631. The signs $+$ and $-$ occur in a 15th century manuscript discovered by Gerhardt at Vienna. The notations $a-b$ and $\frac{a}{b}$ for the fraction were adopted from the Arabians.

canonical equations raised a strong presumption for the soundness of this view of the structure of an equation.

The first statement of the fundamental theorem and of the relations between coefficients and roots occurs in a remarkably clever and modern little book, the *Invention Nouvelle en l'Algebre*, of *Albert Girard*, published in Amsterdam in 1629, two years earlier, therefore, than the *Artis Anatyticae Praxis*. Girard stands in no fear of imaginary roots, but rather insists on the wisdom of recognizing them. They never occur, he says, except when real roots are lacking, and then in number just sufficient to fill out the entire number of roots to equality with the degree of the equation.

Girard also anticipated Descartes in the geometrical interpretation of negatives. But the *Invention Nouvelle* does not seem to have attracted much notice, and the genius and authority of Descartes were needed to give the interpretation general currency.

15. ACCEPTANCE OF THE NEGATIVE, THE GENERAL IRRATIONAL, AND THE IMAGINARY AS NUMBERS.

118. Descartes' Géométrie and the Negative. The *Géométrie* of Descartes appeared in 1637. This famous little treatise enriched geometry with a general and at the same time simple and natural method of investigation: the method of representing a geometric curve by an equation, which, as Descartes puts it, expresses generally the relation of its points to those of some chosen line of reference.[1] To form such equations Descartes represents line segments by letters,—the known by a, b, c, etc., the unknown by x and y. He supposes a perpendicular, y, to be dropped from any point of the curve to the line of reference, and then the equation to be found from the known properties of the curve which connects y with x, the distance of y from a fixed point of the line of reference. This is the equation of the curve in that it is satisfied by the x and y of each and every curve-point.[2] To meet the difficulty that the mere length of the perpendicular (y) from a curve-point will not indicate to which side of the line of reference the point lies, Descartes makes the convention that perpendiculars on opposite sides of this line (and similarly intercepts (x) on opposite sides of the point of reference) shall have opposite algebraic signs.

This convention gave the negative a new position in mathematics. Not only was a "real" interpretation here found for it, the lack of which had made its position so difficult hitherto, but it was made indispensable, placed on a footing of equality with the positive. The acceptance of the negative in algebra kept pace with the spread of Descartes' analytical method in geometry.

119. Descartes' Geometric Algebra. But the *Géométrie* has another and perhaps more important claim on the attention of the historian of algebra. The entire method of the book rests on the assumption—made only tacitly, to be sure, and without knowledge of its significance—that two algebras are formally identical whose fundamental operations are formally the same; *i. e.* subject to the same laws of combination.

For the algebra of the *Géométrie* is not, as is commonly said, mere numerical algebra, but what may for want of a better name be called the algebra of line segments. Its symbolism is the same as that of numerical algebra; but symbols which there represent numbers here represent line segments. Not only is this the case with the letters a, b, x, y, etc., which are mere names (*noms*) of line segments, not their numerical measures, but with the algebraic combinations of these letters. $a + b$ and $a - b$ are respectively the sum and difference of the line segments a and b; ab, the fourth proportional to an assumed unit line, a, and b; $\frac{a}{b}$, the fourth proportional to b, a, and the unit line; and $\sqrt{a}$, $\sqrt[3]{a}$, etc., the first, second, etc., mean proportionals to the unit line and a.[3]

Descartes' justification of this use of the symbols of numerical algebra is that the geometric constructions of which he makes $a + b, a - b$, etc., represent the results are "the same" as numerical addition, subtraction, multiplication, division, and evolution,

[1] See Géométrie, Livre II. In Cousin's edition of Descartes' works, Vol. V, p. 337.

[2] Descartes fails to recognize a number of the conventions of our modern Cartesian geometry. He makes no formal choice of two axes of reference, calls abscissas y and ordinates x, and as frequently regards as positive ordinates below the axis of abscissas as ordinates above it.

[3] Géométrie, Livre I. Ibid. pp. 313–314.

respectively. Moreover, since all geometric constructions which determine line segments may be resolved into combinations of these constructions as the operations of numerical algebra into the fundamental operations, the correspondence which holds between these fundamental constructions and operations holds equally between the more complex constructions and operations. The entire system of the geometric constructions under consideration may therefore be regarded as formally identical with the system of algebraic operations, and be represented by the same symbolism.

In what sense his fundamental constructions are "the same" as the fundamental operations of arithmetic, Descartes does not explain. The true reason of their formal identity is that both are controlled by the commutative, associative, and distributive laws. Thus in the case of the former as of the latter, $ab = ba$, and $a(bc) = abc$; for the fourth proportional to the unit line, a, and b is the same as the fourth proportional to the unit line, b, and a; and the fourth proportional to the unit line, a, and bc is the same as the fourth proportional to the unit line, ab, and c. But this reason was not within the reach of Descartes, in whose day the fundamental laws of numerical algebra had not yet been discovered.

120. The Continuous Variable. Newton. Euler. It is customary to credit the *Géométrie* with having introduced the *continuous variable* into mathematics, but without sufficient reason. Descartes prepared the way for this concept, but he makes no use of it in the *Géométrie*. The x and y which enter in the equation of a curve he regards not as variables but as indeterminate quantities, a pair of whose values correspond to each curve-point.[4] The real author of this concept is Newton (1642–1727), of whose great invention, the method of fluxions, continuous variation, "flow," is the fundamental idea.

But Newton's calculus, like Descartes' algebra, is geometric rather than purely numerical, and his followers in England, as also, to a less extent, the followers of his great rival, Leibnitz, on the continent, in employing the calculus, for the most part conceive of variables as lines, not numbers. The geometric form again threatened to become paramount in mathematics, and geometry to enchain the new "analysis" as it had formerly enchained the Greek arithmetic. It is the great service of *Euler* (1707–1783) to have broken these fetters once for all, to have accepted the *continuously variable number* in its purity, and therewith to have created the pure analysis. For the relations of continuously variable numbers constitute the field of the pure analysis; its central concept, the *function*, being but a device for representing their interdependence.

121. The General Irrational. While its concern with variables puts analysis in a certain opposition to elementary algebra, concerned as this is with constants, its establishment of the continuously variable number in mathematics brought about a rich addition to the number-system of algebra—the *general irrational*. Hitherto the only irrational numbers had been "surds," impossible roots of rational numbers; henceforth their domain is as wide as that of all possible lines incommensurable with any assumed unit line.

122. The Imaginary, a Recognized Analytical Instrument. Out of the excellent results of the use of the negative grew a spirit of toleration for the imaginary. Increased attention was paid to its properties. Leibnitz noticed the real sum of conjugate imaginaries (1676–7); Demoivre discovered (1730) the famous theorem

$$(\cos\theta + i\sin\theta)^n = \cos n\theta + i\sin n\theta;$$

[4] Géométrie, Livre II. Ibid. pp. 337–338.

and Euler (1748) the equation

$$\cos\theta + i\sin\theta = e^{i\theta},$$

which plays so great a rôle in the modern theory of functions.

Euler also, practising the method of expressing complex numbers in terms of modulus and angle, formed their products, quotients, powers, roots, and logarithms, and by many brilliant discoveries multiplied proofs of the power of the imaginary as an analytical instrument.

123. Argand's Geometric Representation of the Imaginary. But the imaginary was never regarded as anything better than an algebraic fiction—to be avoided, where possible, by the mathematician who prized purity of method—until a method was discovered for representing it geometrically. A Norwegian, *Wessel*,[5] published such a method in 1797, and a Frenchman, *Argand*, the same method independently in 1806.

As +1 and -1 may be represented by unit lines drawn in opposite directions from any point, O, and as i (*i. e.* $\sqrt{-1}$) is a mean proportional to +1 and -1, it occurred to Argand to represent this symbol by the line whose direction with respect to the line +1 is the same as the direction of the line -1 with respect to it; viz., the unit perpendicular through O to the 1-line. Let only the *direction* of the 1-line be fixed, the position of the point O in the plane is altogether indifferent.

Between the segments of a given line, whether taken in the same or opposite directions, the equation holds:

$$AB + BC = AC.$$

It means nothing more, however, when the directions of AB and BC are opposite, than that the result of carrying a moving point from A first to B, and thence back to C, is the same as carrying it from A direct to C. But in this sense the equation holds equally when A, B, C are not in the same right line.

Given, therefore, a complex number, $a + ib$; choose any point A in the plane; from it draw a line AB, of length a, in the direction of the 1-line, and from B a line BC, of length b, in the direction of the i-line. The line AC, thus fixed in length and direction, but situated anywhere in the plane, is Argand's picture of $a + ib$.

Argand's skill in the use of his new device was equal to the discovery of the demonstration given in §54, that every algebraic equation has a root.

124. Gauss. The Complex Number. The method of representing complex numbers in common use to-day, that described in §42, is due to Gauss. He was already in possession of it in 1811, though he published no account of it until 1831.

To Gauss belongs the conception of i as an independent unit co-ordinate with 1, and of $a + ib$ as a *complex* number, a sum of multiples of the units 1 and i; his also is the name "complex number" and the concept of complex numbers in general, whereby $a + ib$ secures a footing in the theory of numbers as well as in algebra.

He too, and not Argand, must be credited with really breaking down the opposition of mathematicians to the imaginary. Argand's *Essai* was little noticed when it appeared, and soon forgotten; but there was no withstanding the great authority of Gauss, and his precise and masterly presentation of this doctrine.[6]

---

[5]See W. W. Beman in Proceedings of the American Association for the Advancement of Science, 1897.

[6]See Gauss, Complete Works, II, p. 174.

16. RECOGNITION OF THE PURELY SYMBOLIC CHARACTER OF ALGEBRA. QUARTERNIONS. AUSDEHNUNGSLEHRE.

125. The Principle of Permanence. Thus, one after another, the fraction, irrational, negative, and imaginary, gained entrance to the number-system of algebra. Not one of them was accepted until its correspondence to some actually existing thing had been shown, the fraction and irrational, which originated in relations among actually existing things, naturally making good their position earlier than the negative and imaginary, which grew immediately out of the equation, and for which a "real" interpretation had to be sought.

Inasmuch as this correspondence of the artificial numbers to things extra-arithmetical, though most interesting and the reason of the practical usefulness of these numbers, has not the least bearing on the nature of their position in *pure* arithmetic or algebra; after all of them had been accepted as numbers, the necessity remained of justifying this acceptance by purely algebraic considerations. This was first accomplished, though incompletely, by the English mathematician, *Peacock.*[1]

Peacock begins with a valuable distinction between *arithmetical* and *symbolical* algebra. Letters are employed in the former, but only to represent positive integers and fractions, subtraction being limited, as in ordinary arithmetic, to the case where subtrahend is less than minuend. In the latter, on the other hand, the symbols are left altogether general, untrammelled at the outset with any particular meanings whatsoever.

It is then *assumed* that the rules of operation applying to the symbols of arithmetical algebra apply without alteration in symbolical algebra; *the meanings of the operations themselves and their results being derived from these rules of operation.*

This assumption Peacock names the *Principle of Permanence of Equivalent Forms*, and illustrates its use as follows:[2]

In arithmetical algebra, when $a > b, c > d$, it may readily be demonstrated that

$$(a-b)(c-d) = ac - ad - bc + bd.$$

By the principle of permanence, it follows that

$$(0-b)(0-d) = 0 \times 0 - 0 \times d - b \times 0 + bd, \text{or} (-b)(-d) = bd.$$

Or again. In arithmetical algebra $a^m a^n = a^{m+n}$, when m and n are positive integers. Applying the principle of permanence,

$$\begin{aligned}(a^{\frac{p}{q}})^q &= a^{\frac{p}{q}} \cdot a^{\frac{p}{q}} \cdots \text{to} q \text{factors}\\ &= a^{\frac{p}{q}+\frac{p}{q}+\cdots \text{to} q \text{terms}}\\ &= a^p,\\ \text{whence} \quad a^{\frac{p}{q}} &= \sqrt[q]{a^p}.\end{aligned}$$

Here the meanings of the product $(-b)(-d)$ and of the symbol $a^{\frac{p}{q}}$ are both derived from certain rules of operation in arithmetical algebra.

[1] Arithmetical and Symbolical Algebra, 1830 and 1845; especially the later edition. Also British Association Reports, 1833.

[2] Algebra, edition of 1845, §§ 631, 569, 639.

Peacock notices that the symbol = also has a wider meaning in symbolical than in arithmetical algebra; for in the former = means that "the expression which exists on one side of it is the result of an operation which is indicated on the other side of it and not performed."[3]

He also points out that the terms "real" and "imaginary" or "impossible" are relative, depending solely on the meanings attaching to the symbols in any particular application of algebra. For a quantity is real when it can be shown to correspond to any real or possible existence; otherwise it is imaginary.[4] The solution of the problem: to divide a group of 5 men into 3 equal groups, is imaginary though a positive fraction, while in Argand's geometry the so-called imaginary is real.

The principle of permanence is a fine statement of the assumption on which the reckoning with artificial numbers depends, and the statement of the nature of this dependence is excellent. Regarded as an attempt at a complete presentation of the doctrine of artificial numbers, however, Peacock's Algebra is at fault in classing the positive fraction with the positive integer and not with the negative and imaginary, where it belongs, in ignoring the most difficult of all artificial numbers, the irrational, in not defining artificial numbers as symbolic results of operations, but principally in not subjecting the operations themselves to a final analysis.

126. The Fundamental laws of Algebra. "Symbolical Algebras." Of the fundamental laws to which this analysis leads, two, the commutative and distributive, had been noticed years before Peacock by the inventors of symbolic methods in the differential and integral calculus as being common to number and the operation of differentiation. In fact, one of these mathematicians, *Servois*,[5] introduced the names *commutative* and *distributive*.

Moreover, Peacock's contemporary, *Gregory*, in a paper "On the Real Nature of Symbolical Algebra," which appeared in the interim between the two editions of Peacock's Algebra,[6] had restated these two laws, and had made their significance very clear.

To Gregory the formal identity of complex operations with the differential operator and the operations of numerical algebra suggested the comprehensive notion of algebra embodied in his fine definition: "symbolical algebra is the science which treats of the combination of operations defined not by their nature, that is, by what they are or what they do, but by the laws of combination to which they are subject."

This definition recognizes the possibility of an entire class of algebras, each characterized primarily not by its subject-matter, but by *its operations and the formal laws to which they are subject*; and in which the algebra of the complex number $a + ib$ and the system of operations with the differential operator are included, the two (so far as their laws are identical) as one and the same particular case.

So long, however, as no "algebras" existed whose laws differed from those of the algebra of number, this definition had only a speculative value, and the general acceptance of the dictum that the laws regulating its operations constituted the essential

---

[3] Algebra, Appendix, §631.

[4] Ibid. §557.

[5] Gergonne's Annales, 1813. One must go back to Euclid for the earliest known recognition of any of these laws. Euclid demonstrated, of integers (Elements, VII, 16), that $ab = ba$.

[6] In 1838. See The Mathematical Writings of D. F. Gregory, p. 2. Among other writings of this period, which promoted a correct understanding of the artificial numbers, should be mentioned Gregory's interesting paper, "On a Difficulty in the Theory of Algebra," Writings, p. 235, and De Morgan's papers "On the Foundation of Algebra" (1839, 1841; Cambridge Philosophical Transactions, VII).

character of algebra might have been long delayed had not Gregory's paper been quickly followed by the discovery of two "algebras," the *quaternions* of *Hamilton* and the *Ausdehnungslehre* of *Grassmann*, in which one of the laws of the algebra of number, the commutative law for multiplication, had lost its validity.

127. Quaternions. According to his own account of the discovery,[7] Hamilton came upon *quaternions* in a search for a second imaginary unit to correspond to the perpendicular which may be drawn in space to the lines 1 and i.

In pursuance of this idea he formed the expressions, $a + ib + jc, x + iy + jz$, in which a, b, c, x, y, z were supposed to be real numbers, and j the new imaginary unit sought, and set their product

$$(a + ib + jc)(x + iy + jz) = ax - by - cz + i(ay + bx) \ + j(az + cx) + ij(bz + cy).$$

The question then was, what interpretation to give ij. It would not do to set it equal to $a' + ib' + jc'$, for then the theorem that the modulus of a product is equal to the product of the moduli of its factors, which it seemed indispensable to maintain, would lose its validity; unless, indeed, $a' = b' = c' = 0$, and therefore $ij = 0$, a very unnatural supposition, inasmuch as $1i$ is different from 0.

No course was left for destroying the ij term, therefore, but to make its coefficient, $bz+cy$, vanish, which was tantamount to supposing, since b, c, y, z are perfectly general, that $ji = -ij$.

Accepting this hypothesis, *denial of the commutative law* as it was, Hamilton was driven to the conclusion that the system upon which he had fallen contained at least three imaginary units, the third being the product ij. He called this k, took as general complex numbers of the system, $a + ib + jc + kd, x + iy + jz + kw$, *quaternions*, built their products, and assuming

$$\begin{aligned} i^2 &= j^2 = k^2 = -1 \\ ij &= -ji = k \\ jk &= -kj = i \\ ki &= -ik = j, \end{aligned}$$

found that the modulus law was fulfilled.

A geometrical interpretation was found for the "*imaginary triplet*" $ib+jc+kd$, by making its coefficients, b, c, d, the rectangular co-ordinates of a point in space; the line drawn to this point from the origin picturing the triplet by its length and direction. Such directed lines Hamilton named *vectors*.

To interpret geometrically the multiplication of i into j, it was then only necessary to conceive of the j axis as rigidly connected with the i axis, and *turned by it* through a right angle in the jk plane, into coincidence with the k axis. The geometrical meanings of other operations followed readily.

In a second paper, published in the same volume of the Philosophical Magazine, Hamilton compares in detail the laws of operation in *quaternions* and the algebra of number, for the first time explicitly stating and naming the *associative* law.

128. Grassmann's Ausdehnungslehre. In the *Ausdehnungslehre*, as Grassmann first presented it, the elementary magnitudes are vectors.

---

[7]Philosophical Magazine, II, Vol. 25, 1844.

The fact that the equation $AB + BC = AC$ always holds among the segments of a line, when account is taken of their directions as well as their lengths, suggested the probable usefulness of directed lengths in general, and led Grassmann, like Argand, to make trial of this definition of addition for the general case of three points, A, B, C, not in the same right line.

But the outcome was not great until he added to this his definition of the product of two vectors. He took as the product ab, of two vectors, a and b, the parallelogram generated by a when its initial point is carried along b from initial to final extremity.

This definition makes a product vanish not only when one of the vector factors vanishes, but also when the two are parallel. It clearly conforms to the distributive law. On the other hand, since

$$(a + b)(a + b) = aa + ab + ba + bb,$$

and $$(a + b)(a + b) = aa = bb = 0,$$

$$ab + ba = 0, \text{ or } ba = -ab,$$

the commutative law for multiplication has lost its validity, and, as in quaternions, an interchange of factors brings about a change in the sign of the product.

The opening chapter of Grassmann's first treatise on the *Ausdehnungslehre* (1844) presents with admirable clearness and from the general standpoint of what he calls "Formenlehre" (the doctrine of forms), the fundamental laws to which operations are subject as well in the *Ausdehnungslehre* as in common algebra.

129. The Doctrine of the Artificial Numbers fully Developed. The discovery of quaternions and the *Ausdehnungslehre* made the algebra of number in reality what Gregory's definition had made it in theory, no longer the sole algebra, but merely one of a class of algebras. A higher standpoint was created, from which the laws of this algebra could be seen in proper perspective. Which of these laws were distinctive, and what was the significance of each, came out clearly enough when numerical algebra could be compared with other algebras whose characteristic laws were not the same as its characteristic laws.

The doctrine of the artificial numbers regarded from this point of view—as symbolic results of the operations which the fundamental laws of algebra define—was fully presented for the negative, fraction, and imaginary, by *Hankel*, in his *Complexe Zahlensystemen* (1867). Hankel re-announced Peacock's principle of permanence. The doctrine of the irrational now accepted by mathematicians is due to *Weierstrass* and *G. Cantor* and *Dedekind.*[8]

A number of interesting contributions to the literature of the subject have been made recently; among them a paper[9] by Kronecker in which methods are proposed for avoiding the artificial numbers by the use of congruences and "indeterminates," and papers[10] by Weierstrass, Dedekind, Hölder, Study, Scheffer, and Schur, all relating to the theory of general complex numbers built from n fundamental units (see page 40).

---

[8]See Cantor in Mathematische Annalen, V, p. 123, XXI, p. 567. The first paper was written in 1871. In the second, Cantor compares his theory with that of Weierstrass, and also with the theory proposed by Dedekind in his *Stetigkeit und irrationals Zahlen* (1872).

The theory of the irrational, set forth in Chapter IV of the first part of this book, is Cantor's.

[9]Journal für die reine und angewandte Mathematik, Vol. 101, p. 337.

[10]Göttinger Nachrichten for 1884, p. 395; 1885, p. 141; 1889, p. 34, p. 237. Leipziger Berichte for 1889, p. 177, p. 290, p. 400. Mathemathische Annalen, XXXIII, p. 49.

SUPPLEMENTARY NOTE, 1902. An elaborate and profound analysis of the number-concept from the ordinal point of view is made by Dedekind in his *Was sind und was sollen die Zahlen?* (1887). This essay, together with that on irrational numbers cited above, has been translated by W. W. Beman, and published by the Open Court Company, Chicago, 1901.

The same point of view is taken by Kronecker in the memoir above mentioned, and by Helmholtz in his *Zählen und Messen* (Zeller-Jubeläum, 1887).

G. Cantor discusses the general notion of cardinal number, and extends it to infinite groups and assemblages in his now famous Memoirs on the theory of infinite assemblages. See particularly Mathematische Annalen, XLVI, p. 489.

Very recently much attention has been given to the question: What is the *simplest* system of consistent and independent laws—or "axioms," as they are called—by which the fundamental operations of the ordinary algebra may be defined? A very complete *résumé* of the literature may be found in a paper by O. Hölder in Leipziger Berichte, 1901. See also E. V. Huntington in Transactions of the American Mathematical Society, Vol. III, p. 264.

9 789393 971661

Printed at Repro India Ltd.